To
Jetty
Annemarie
and Frank

CHEMICALS FROM SYNTHESIS GAS

CATALYSIS BY METAL COMPLEXES

ROGER A. SHELDON

Research and Development Department, Océ Andeno BV
Venlo, The Netherlands

CHEMICALS FROM SYNTHESIS GAS

Catalytic Reactions of CO and H$_2$

D. REIDEL PUBLISHING COMPANY

A MEMBER OF THE KLUWER ACADEMIC PUBLISHERS GROUP

DORDRECHT / BOSTON / LANCASTER

Library of Congress Cataloging in Publication Data

Sheldon, Roger A.
    Chemicals from synthesis gas.

(Catalysis by metal complexes)
Includes bibliographical references and index.
1. Feedstock. 2. Synthesis gas. 3. Carbon monoxide.
4. Hydrogen. I. Title. II. Series.
TP201.S44   1983   661'.804   83-4428
ISBN 90-277-1489-4

Published by D. Reidel Publishing Company,
P.O. Box 17, 3300 AA Dordrecht, Holland.

Sold and distributed in the U.S.A. and Canada
by Kluwer Boston Inc.,
190 Old Derby Street, Hingham, MA 02043, U.S.A.

In all other countries, sold and distributed
by Kluwer Academic Publishers Group,
P.O. Box 322, 3300 AH Dordrecht, Holland.

CONTENTS

PREFACE

The origins of the petrochemical industry can be traced back to the 1920s when simple organic chemicals such as ethanol and isopropanol were first prepared on an industrial scale from by-products (ethylene and propylene) of oil refining. This oil-based petrochemical industry, with lower olefins and aromatics as the key building blocks, rapidly developed into the enormous industry it is today. A multitude of products that are indispensible to modern day society, from plastics to pharmaceuticals, are derived from oil and natural gas-based hydrocarbons.

The industry had its heyday in the '50s and '60s when predictions of future growth rates tended to be exponential curves. However, two developments that took place in the early '70s disturbed this simplistic and optimistic view of the future. Firstly, the publication of the report for the Club of Rome on the 'Limits to Growth' emphasized the finite nature of non-renewable fossil fuel resources. Secondly, the Oil Crisis of 1973 emphasized the vulnerability of an energy and chemicals industry that is based largely on a single raw material.

These developments stimulated, almost overnight, a search for alternative sources of energy and raw materials, such as coal, oil shale, tar sands, biomass, etc. Careful inspection of the various options, however, reveals that there is no simple, all-encompassing solution. In the words of H. L. Mencken: "for every human problem there is a solution that is simple, neat and wrong." Agricultural production of biomass, for example, may constitute an attractive alternative for a country, such as Brazil, with large areas of land mass, but not be viable for Western European countries.

Synthesis gas (mixtures of carbon monoxide and hydrogen) can play a key role in the solution of this predicament concerning the future availability of feedstocks for hydrocarbon fuels and organic chemicals. Syn gas has the important advantage that it can be produced from essentially any carbon source, from methane to manure. Thus, a chemicals industry based on syn gas would be largely independent of the ultimate carbon source, whether it be natural gas, coal, household garbage or agricultural wastes.

Potential applications of syn gas include conversion to liquid fuels and the production of a wide range of industrial organic chemicals. Some of these chemicals, such as methanol and acetic acid, are already produced industrially from syn gas whilst many others stand at the threshold of commercialization.

xi

One potentially attractive approach, explored extensively in this book, is to use syn gas-derived methanol as a feedstock for a wide range of organic chemicals. Methanol has the advantage that it can be readily transported using existing infrastructures. Moreover, it is a dual purpose raw material, suitable as both a fuel and a chemical feedstock.

The principle aim of this book is to provide an overview of the reactions of syn gas, many of which have been developed in the last few years. The subject matter is treated primarily from the point of view of chemicals manufacture. In the treatment of the conversion of syn gas to hydrocarbons, for example, emphasis is placed on the production of hydrocarbon base chemicals rather than liquid fuels. It is recognized, however, that the production of liquid fuels and hydrocarbon feedstocks are, just as in the oil-based industry, intricately related.

The potential applications of syn gas are not limited to the production of bulk organic chemicals. Reactions of syn gas or carbon monoxide alone can be effectively utilized in the synthesis of a wide variety of fine organic chemicals. Many of these processes represent attractive alternatives to more conventional synthetic methods, but they have yet to be "discovered" by the majority of organic chemists.

The chemistry of syn gas is the chemistry of organotransition metal complexes, and in particular of metal carbonyls and related complexes. Virtually none of the reactions discussed in this book proceed in the absence of transition metal catalysts. An understanding of the fundamental steps involved in the transition metal catalysis of the reactions of CO and H_2 is of paramount importance for understanding the reactions of syn gas. Much progress has been booked in the last decade, particularly with respect to the mechanisms of homogeneous catalysis, and mechanistic aspects are reviewed in depth before going on to discuss the various reactions in detail. Although both homogenous and heterogeneous catalysis of syn gas conversions are covered there is a bias towards the former, mainly because recent developments largely involve homogeneous catalysis, certainly in the context of chemicals manufacture.

This book is directed towards chemists engaged in organic synthesis, organotransition metal chemistry and catalysis, both in industrial laboratories and academic institutions. A primary aim is to cultivate a deeper understanding and stimulate a wider utilization of transition metal catalysis in organic synthesis. The book may be used to augment a more traditional textbook on organic synthesis, particularly in the context of industrial applications. It could also be used for a special topics course on modern aspects of catalysis or of the petrochemical industry. A chemical background beyond the first-year courses in organic and inorganic chemistry is desirable, since a rudimentary knowledge of the kinetics and mechanism of organic and inorganic reactions is assumed.

The literature is covered through the end of 1981, drawing extensively from the patent literature in addition to the standard journals.

Finally, I wish to acknowledge the important contribution of my wife, Jetty, who typed the entire manuscript and without whose constant encouragement and understanding this book would not have been possible. All mistakes are, however, the responsibility of the author alone.

Roger A. Sheldon

Venray, Holland
June, 1982.

GLOSSARY

Alkoxycarbonylation: Reaction of a substrate (olefin, acetylene, alkyl halide, etc.) with a mixture of carbon monoxide and an alcohol to give a carboxylic acid ester, e.g.

$$RCH{=}CH_2 + CO + R'OH \longrightarrow RCH_2CH_2CO_2R'$$

Alpha olefins: Olefins containing the double bond at the terminal position in the carbon chain, e.g. 1-octene. An alternative name in common usage is *terminal olefin*.

Amidocarbonylation: Reaction of aldehydes with a mixture of carbon monoxide and an amide to give an amino acid derivative, e.g.

$$RCHO + CO + R'CONH_2 \longrightarrow RCH(NHCOR')CO_2H$$

The aldehyde can be generated *in situ* by, for example, olefin hydroformylation.

Aminomethylation: see *Hydroformamination*.

Carbonylation: Generic term for reactions of substrates with carbon monoxide as opposed to reactions with syn gas.

Carbonylmetallate anion: The anion derived from dissociation of a metal carbonyl hydride, e.g.

$$HCo(CO)_4 \rightleftharpoons H^+ + Co(CO)_4^-$$

Catalyst: An additive that increases the rate of a chemical process without itself being consumed. The actual catalytic species may differ from the compound added to the reaction mixture (i.e. the catalyst precursor).

Conversion: Amount of reactant consumed in a chemical process, expressed as a percentage of the amount of reactant charged.

Coordinative unsaturation: A metal complex is coordinatively unsaturated when it does not contain its full complement of ligands required for a stable, inert gas configuration (see electron rule), e.g.

$$HCo(CO)_4 \rightleftharpoons HCo(CO)_3 + CO$$

coordinatively structured	coordinatively unstructured
18e	16e

Electron rule: Predicts that metal complexes possessing an 18-electron (inert gas) configuration should be stable. Also referred to as the *inert gas rule*. For the purpose of electron counting covalent ligands, e.g. halide, hydride, etc., are assigned one electron whilst coordinate ligands, e.g. trialkylphosphines, carbon monoxide, are assigned two electrons, e.g.

$$HCo(CO)_4 = 1 \text{ (hydride)} + 9 \text{ (cobalt)} + 8 \text{ (4CO)} = 18 \text{ electrons}$$

A corollary of the rule is the so-called *16 and 18 electron rule* which predicts that catalytic processes should preferably proceed through steps involving alternating 16 and 18 electron intermediates.

Electron transfer: Addition or removal of an electron from a chemical species resulting in its reduction or oxidation, respectively. Reference to electron transfer may be made in describing the stoichiometry of a redox process or the mechanism of a particular step (usually outer sphere).

Hapto: System of notation used to specify the number of atoms in a ligand that are directly attached to the metal. The number of atoms attached to the metal are specified by a prefix dihapto, trihapto, tetrahapto, etc.. The term is derived from the Greek word, *haptein*, to fasten. In formulae the notation η^n is used to denote the number of atoms bound to the metal, e.g. η^2 = dihapto.

Heterolysis: Fragmentation of a neutral, diamagnetic compound into an ion pair, e.g.,

$$HCo(CO)_4 \longrightarrow H^+ + Co(CO)_4{}^-$$

Homolysis: Fragmentation of a diamagnetic compound into two paramagnetic species, e.g.,

$$Co_2(CO)_8 \longrightarrow 2 Co(CO)_4$$

Homologation: Reaction of an organic substrate with syn gas resulting in an extension of the carbon chain by one or more methylene (CH_2) units, e.g. alcohol homologation:

$$ROH + CO + 2 H_2 \longrightarrow RCH_2OH + H_2O$$

β-Hydride elimination: Loss of a hydridometal species from a complex induced by the removal of a β-proton, e.g.

$$M-CH_2CH_3 \longrightarrow M-H + CH_2{=}CH_2$$

Hydrocarbonylation: Generic term for the reaction of a substrate with syn gas as opposed to reactions with carbon monoxide alone.

Hydroformamination: Reaction of an olefin with a mixture of syn gas and an amine, resulting in alkylation of the latter according to the stoichiometry:

$$RCH{=}CH_2 + CO + 2\,H_2 + R_2NH \longrightarrow R(CH_2)_3NR_2 + H_2O$$

The reaction is sometimes referred to as *aminomethylation.*

Hydroformylation: Reaction of an olefin with syn gas to afford an aldehyde via the stoichiometry:

$$RCH{=}CH_2 + CO + H_2 \longrightarrow RCH_2CH_2CHO$$

Sometimes the aldehyde is hydrogenated *in situ* to give the corresponding alcohol.

Hydroxycarbonylation: Reaction of a substrate (olefin, acetylene, organic halides, etc.) with a mixture of carbon monoxide and water to give a carboxylic acid, e.g.

$$RCH{=}CH_2 + CO + H_2O \longrightarrow RCH_2CH_2CO_2H$$

The reaction is sometimes referred to as *hydrocarboxylation* although this term is a less accurate description of the reaction.

Internal olefins: Olefins containing an internal, i.e. non-terminal, double bond, e.g. 2-octene.

Linearity: Refers to the percentage of linear isomer formed in hydroxy-carbonylations, hydroformylations, etc. where the reaction affords both linear and branched isomers.

Migratory insertion: The insertion of a coordinated ligand into a metal-ligand covalent bond in the same metal complex, e.g.

$$\begin{array}{c} M{-}R \\ | \\ CO \end{array} \rightleftharpoons M{-}COR$$

The reverse process is *migratory elimination.*

Oxidation number: The formal oxidation state of the metal centre in a complex determined as:

metal charge = charge on complex ion − ligand charges.

The designation is arbitrary but is useful as a bookkeeping device to keep track of electron changes in redox processes.

Oxidative addition: Reaction in which the oxidation of a metal complex by an electrophile is accompanied by an increase in its coordination number, e.g.

$$M^0 + R{-}Br \rightleftharpoons R{-}M^{II}{-}Br$$

The term is usually applied to two-equivalent changes of the metal centre without regard to the mechanism. The reverse process is *reductive elimination*.

Oxygenates: General term used to describe oxygen-containing organic substrates, e.g. C_2 oxygenates comprise ethanol, acetaldehyde, acetic acid, ethylene glycol, etc.

Promotor: An additive that increases the rate of a catalytic process. Also referred to as a *co-catalyst*, e.g. iodide ion is a promotor or co-catalyst in rhodium-catalyzed alcohol carbonylations.

Reductive elimination: The reverse of oxidative addition.

Selectivity: The amount of a particular product formed divided by the amount of the reactant consumed, expressed as a percentage.

Synthesis gas, Syn gas: Refers to mixtures of carbon monoxide and hydrogen of varying composition.

Terminal olefins: see *alpha olefins*.

Water Gas Shift Reaction: Reaction of carbon monoxide with water to afford a mixture of carbon dioxide and hydrogen:

$$CO + H_2O \rightleftharpoons CO_2 + H_2$$

Yield: The amount of a particular product formed divided by the amount of reactant charged, expressed as a percentage. The *yield* equals the *conversion* times the *selectivity*.

ABBREVIATIONS

acac	acetylacetonate
Am	amyl (pentyl)
Ar	aryl
BTX	benzene, toluene and xylene
Bu	butyl
ee	enantiomeric excess
Et	ethyl
FT	Fischer-Tropsch process
L	neutral ligand (usually phosphine)
M	metal
Me	methyl
MTBE	methyl *tert*-butyl ether
OAc	acetate
P	polymeric group
Ph	phenyl
Pr	propyl
R	alkyl
SET	single electron transfer
WGSR	water gas shift reaction
X	anionic ligand (usually halide)

CHAPTER 1

INTRODUCTION TO PETROCHEMICALS

1.1 Background

Synthesis gas, commonly abbreviated to *syn gas* is the trivial name for mixtures of carbon monoxide and hydrogen. The CO/H_2 molar ratio varies from *ca.* 2 : 1 to 1 : 3 depending on the method used for generating the syn gas (see below). It can be produced from almost anything containing carbon and hydrogen, from methane to manure [1]. The most important source of syn gas at the present time is steam reforming of methane (natural gas) which affords CO/H_2 in a molar ratio of 1 : 3 (see Section 1.5).

Syn gas plays an important role in the chemical industry as a source of pure hydrogen and pure carbon monoxide. The hydrogen is consumed in large quantities for the production of ammonia by reaction with atmospheric nitrogen in the Haber process. Syn gas is also used as such in, for example, the production of methanol (Reaction 1):

$$CO + 2 H_2 \longrightarrow CH_3OH \tag{1}$$

Methanol is, in turn, a precursor of a host of other important chemicals such as formaldehyde, acetic acid, methyl chloride and methylamines. Thus, syn gas has for a long time played an important role in the chemical industry. However, the last five years or more have witnessed a burgeoning interest in the chemistry of syn gas, so-called C_1 chemistry. This interest has been stimulated by the prospect that syn gas will, in the future, become available on a large scale, at low cost from coal gasification. The exponential increase in the price of crude oil since the Oil Crisis of 1973 has provided an ever-growing incentive for oil companies to consider alternative sources of liquid fuels and petrochemicals. In particular coal, the known reserves of which are more than ten times those of oil, has received much attention. There is good reason to believe, therefore, that coal-based syn gas, will in the not too distant future, play an important role as a basic raw material for the chemical industry.

Before going on to consider the reactions of syn gas we shall first take a look at the conventional routes for producing the major petrochemicals. This will set

1

the scene for comparison with alternative, syn gas based routes to these chemicals which is the subject of this book.

1.2 Sources of Hydrocarbon Building Blocks

Natural gas and crude oil, or petroleum, presently account for about 90% of the organic chemicals produced in the world. They are the primary sources of the seven basic building blocks—ethylene, propylene, butadiene, benzene, toluene, xylenes and methanol—on which a vast organic chemical industry is based. Table 1.I shows the volumes manufactured in the United States in 1979 of these key basic petrochemicals. The rank order would be essentially the same in any developed country. By far the most important base chemical, US production of which was *ca.* 14 million tonnes in 1979, is ethylene. It is the raw material for roughly half (on a weight basis) of all the organic chemicals produced.

TABLE 1.I

Production of basic petrochemicals in the United States in 1979[a]

Chemical	Production (millions of tonnes)
Ethylene	13.6
Propylene	6.44
Benzene	5.57
Toluene	3.32
Methanol	3.34
Butadiene	1.62
p-Xylene	2.11
o-Xylene	0.46

[a] Taken from *C. and E. News*, Dec. 22, 1980, p. 32.

It should be emphasized, however, that liquid fuels are the most important products of an oil refinery and that less than 10% of a barrel of crude oil is used for chemicals manufacture. The petrochemical industry was originally developed as an outlet for ethylene and propylene-containing refinery gases, the byproducts of oil processing. Thus, the origins of the petrochemical industry can be traced back to the work of Carleton Ellis [2] who, in 1920, developed a process for the manufacture of isopropanol, the first petrochemical, by passing refinery gases into concentrated sulfuric acid followed by subsequent hydrolysis.

$$CH_3CH{=}CH_2 + H_2SO_4 \longrightarrow CH_3CH(OSO_3H)CH_3 \xrightarrow{H_2O}$$
$$CH_3CH(OH)CH_3 + H_2SO_4 \qquad (2)$$

In order to understand how petrochemicals are derived from crude oil we must first consider what happens in a modern oil refinery. The crude oil is separated by atmospheric and vacuum distillation into various fractions (see Table 1.II). The lowest boiling fraction consists of methane, ethane, propane and butane and is similar to natural gas. It is, in principle, useful for both fuels and chemicals but in practice much of it is flared because of the expense of recovery. The second and third fractions are called light naphtha (b.p. 20–150°C) and heavy naphtha (b.p. 150–200°C), respectively, and contain predominantly C_5-C_{10} alkanes and cycloalkanes. Naphtha is a major source of both gasoline and chemicals. The higher boiling fractions – kerosene, gas oil and lubricating oil – are generally less suitable than naphtha as sources of chemicals although they can be cracked to light olefins and gasoline-range hydrocarbons (see below).

TABLE 1.II
Crude oil distillation fractions

Fraction	Boiling range (°C)	Composition / use
Gases	< 25	C_1-C_4 alkanes
Light naphtha	20–150	Mainly C_5-C_{10} alkanes
Heavy naphtha	150–200	and cycloalkanes. Useful for both fuel and chemicals.
Kerosene	175–275	C_9-C_{16} compounds useful for jet, tractor and heating fuel.
Gas oil	200–400	$C_{15}-C_{25}$ compounds useful for diesel and heating fuel.
Lubricating oil/ Heavy fuel oil	350	$C_{20}-C_{70}$. Used for lubrication and boiler fuel.
Asphalt	residue	Used in structural applications.

Two reactions – steam cracking and catalytic reforming – account for the production of most of the base chemicals for the petrochemical industry. Thus, the three most important olefins – ethylene, propylene and butadiene – are coproduced in various ratios by the thermal cracking, at 750–850°C in the presence of steam, of feedstocks ranging from ethane to crude oil (see Table 1.III). In Europe about 90% of the ethylene is produced by naphtha cracking. In the United States, in contrast, the demand for gasoline is much higher and ethylene is produced mainly by ethane cracking. As feedstocks progress from ethane to heavier fractions with a lower hydrogen content the yield of ethylene decreases and the feed required per pound of ethylene produced increases significantly.

The major aromatic feedstocks – benzene, toluene and xylenes (BTX) – are obtained as byproducts of olefin production by naphtha cracking. They are also derived from the catalytic reforming of naphtha, which involves the dehydrocyclization of alkanes over supported platinum or rhenium catalysts.

TABLE 1.III

Typical yields from the steam cracking of various feedstocks[a]

Feedstock	Products (wt%)				
	Ethylene	Propylene	Butadiene	BTX	Other
Ethane	84.0	1.4	1.4	0.4	12.8
Propane	44.0	15.6	3.4	2.8	34.2
Light naphtha	40.3	15.8	4.9	4.8	34.2
Full range naphtha	31.7	13.0	4.7	13.7	36.9
Light gas oil	25.0	12.4	4.8	11.2	46.6
Crude oil	32.8	4.4	3.0	14.4	45.4

[a] Data taken from L.F. Hatch and S. Matar, *From Hydrocarbons to Petrochemicals*, Gulf Publishing Co., Houston, 1981, p. 71.

The sources of the major hydrocarbon feedstocks are summarized in Figure 1.1.

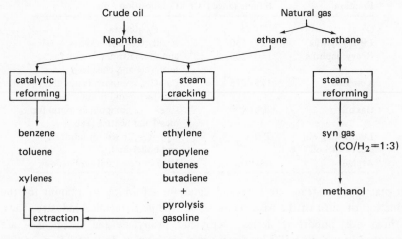

Figure 1.1

Now that we have seen how the seven base chemicals are produced from crude oil or natural gas we can examine the wide variety of organic chemicals which they are subsequently converted to.

1.3 Conversion to Industrial Derivatives

CHEMICALS FROM ETHYLENE

About half of the ethylene produced is used for the production of polyethylene.

Some is converted, in the presence of a triethylaluminium catalyst, to ethylene oligomers which are important raw materials for the manufacture of detergents. The rest is mainly converted, by reaction with cheap materials such as oxygen, water, chlorine and syn gas, into a variety of industrially important derivatives (see Figure 1.2). It is noteworthy that the majority of these processes involve the introduction of oxygen into the molecule.

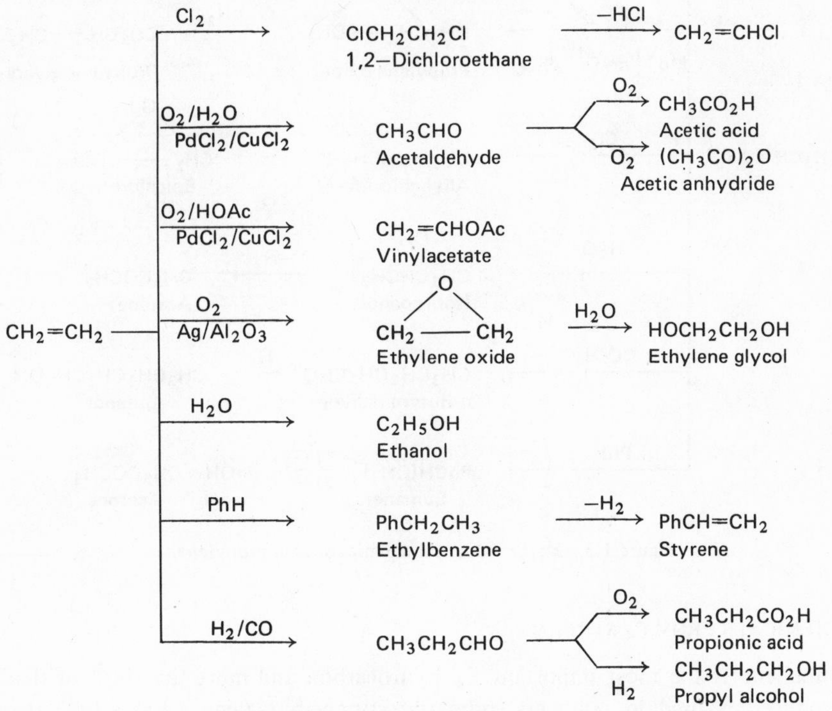

Figure 1.2 Major industrial chemicals from ethylene.

CHEMICALS FROM PROPYLENE

After ethylene the most important base chemical is propylene. In contrast to ethylene, propylene also has non-chemical uses — about half the propylene produced is used in the manufacture of alkylates for gasoline. In common with ethylene, propylene and its derivatives constitute the raw materials for a variety of polymers, such as polypropylene, polyurethanes and epoxy resins. The major industrial derivatives of propylene, virtually all of which involve the introduction of oxygen into the molecule, are produced by the routes outlined in Figure 1.3.

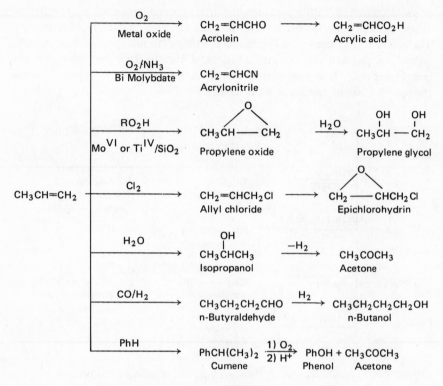

Figure 1.3. Major industrial chemicals from propylene.

CHEMICALS FROM C_4 STREAMS

Butadiene is the most important C_4 hydrocarbon and more than 90% of that produced is used in polymers, primarily styrene-butadiene rubbers for automobile tyres. The major non-polymer application of butadiene is probably in the manufacture of hexamethylene diamine via the nickel (0)-catalysed addition of HCN and hydrogenation of the intermediate adiponitrile [3].

$$CH_2=CH-CH=CH_2 + 2HCN \longrightarrow NC(CH_2)_4CN$$
$$\xrightarrow{H_2} H_2N(CH_2)_6NH_2 \tag{3}$$

Small amounts of butenes are formed in the steam cracking of naphtha as byproducts of ethylene manufacture. The n-butenes, 1-butene and 2-butenes, are raw materials for the manufacture of sec-butanol and methyl ethyl ketone, e.g.

$$CH_3 CH_2 CH = CH_2 \quad \xrightarrow[\text{H}^+]{\text{H}_2\text{O}} \quad CH_3 CH_2 CH(OH)CH_3$$
$$\downarrow -H_2$$
$$\xrightarrow[[PdCl_2/CuCl_2]]{O_2/H_2O} \quad CH_3 CH_2 COCH_3$$

Butenes and *n*-butane are also used as raw materials for acetic acid manufacture by catalytic oxidation (see Chapter 5 for a discussion of acetic acid manufacture).

The single largest use of isobutene is in the production (reaction 4) of methyl *tert*-butyl ether (MTBE), which is added to gasoline as an octane booster [4]. Isobutene is also a potential raw material for methacrolein and methacrylonitrile production via reactions analogous to the conversion of propylene into acrolein and acrylonitrile, respectively (see Figure 1.3).

$$(CH_3)_2 C = CH_2 + CH_3 OH \quad \xrightarrow{\text{H}^+} \quad (CH_3)_3 COCH_3 \qquad (4)$$
$$\text{MTBE}$$

CHEMICALS FROM AROMATIC HYDROCARBONS (BTX)

Steam cracking of naphtha produces substantial amounts of a so-called pyrolysis gasoline fraction which contains benzene as the main component, together with smaller amounts of toluene and xylenes. Catalytic reforming, on the other hand, typically gives a BTX mixture containing 50% toluene, 35–45% xylenes and 10–15% benzene. Since benzene has the highest demand much of the toluene produced in catalytic reforming is subsequently converted to benzene by dehydromethylation.

The major outlets of benzene are in the manufacture of ethylbenzene and cumene via alkylation with ethylene and propylene, respectively. Cumene is subsequently converted to phenol and acetone via liquid phase oxidation (see Figure 1.3). Ethylbenzene is converted to styrene by dehydrogenation. Alternatively, styrene is produced by dehydration of phenylmethylcarbinol, the co-product of propylene oxide manufacture by ethylbenzene hydroperoxide epoxidation of propylene [5,6].

$$PhCH_2 CH_3 \xrightarrow{O_2} PhCH(CH_3)O_2 H \xrightarrow[Mo^{VI} \text{ or } Ti^{IV}/SiO_2]{CH_3CH=CH_2}$$
$$CH_3 CH \overset{O}{-\!\!-\!\!-} CH_2 + PhCH(CH_3)OH \xrightarrow{-H_2O} PhCH=CH_2$$

Cyclohexane, a raw material for adipic acid and caprolactam, comes third after styrene and cumene as an outlet for benzene (see Scheme 1.1). Adipic acid and caprolactam are, in turn, raw materials for the two important polymers, nylon-6,6 and nylon-6, respectively.

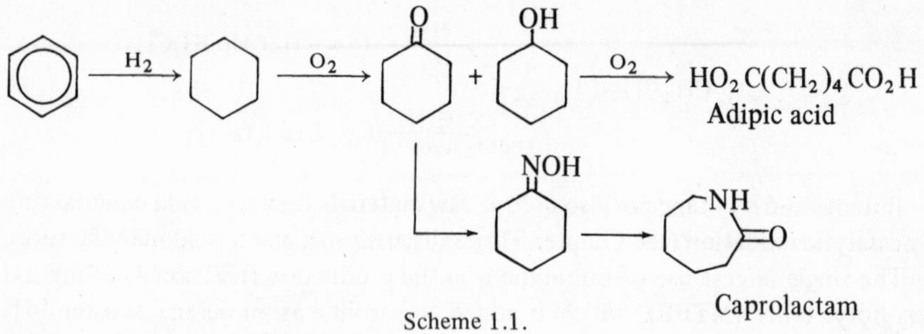

Scheme 1.1.   Caprolactam

Because toluene, like propylene, is abundantly available as a by-product of ethylene manufacture, much effort has been devoted to finding uses for it. As noted above a large part of it is converted to the more desirable benzene. Another approach is to convert toluene directly to established end-products which otherwise are obtained from benzene. For example, toluene can be converted to phenol via benzoic acid [7]. A more recent example that makes use of the olefin metathesis reaction is the two-step conversion of toluene to styrene via Reactions 5 and 6 [8].

$$2 \, PhCH_3 \xrightarrow{PbO} PhCH{=}CHPh \tag{5}$$

$$PhCH{=}CHPh + CH_2{=}CH_2 \xrightarrow{[WO_3/CaO/SiO_2]} 2 \, PhCH{=}CH_2 \tag{6}$$

The xylenes are produced together with ethylbenzene as the C_8-aromatics fraction from catalytic reforming. The *para* isomer is the most important for chemicals manufacture. Virtually all of the *p*-xylene produced is converted to terephthalic acid (and dimethyl terephthalate), via cobalt-catalysed autoxidation (Reaction 7).

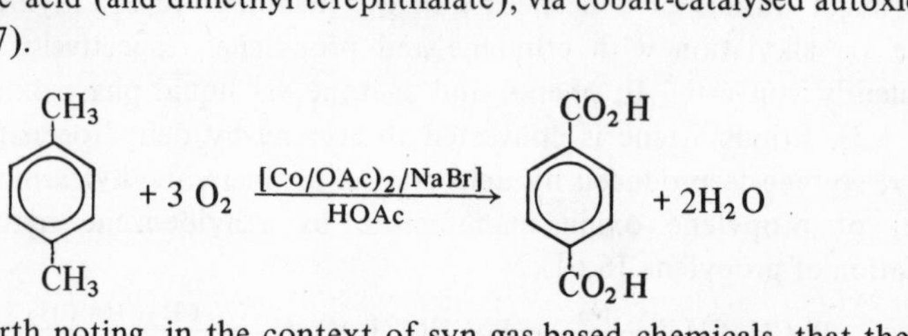

$$\underset{CH_3}{\overset{CH_3}{\bigcirc}} + 3 \, O_2 \xrightarrow[HOAc]{[Co/OAc)_2/NaBr]} \underset{CO_2H}{\overset{CO_2H}{\bigcirc}} + 2H_2O \tag{7}$$

It is worth noting, in the context of syn gas-based chemicals, that the more recently developed route [9], starting from toluene and carbon monoxide (see Reaction 8) may become increasingly competitive with the above route in the future.

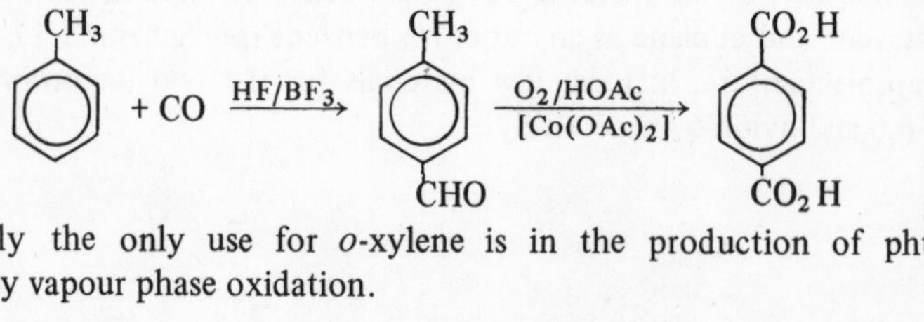

$$\underset{}{\overset{CH_3}{\bigcirc}} + CO \xrightarrow{HF/BF_3} \underset{CHO}{\overset{CH_3}{\bigcirc}} \xrightarrow[{[Co(OAc)_2]}]{O_2/HOAc} \underset{CO_2H}{\overset{CO_2H}{\bigcirc}} \tag{8}$$

Essentially the only use for *o*-xylene is in the production of phthalic anhydride by vapour phase oxidation.

1.4 Chemicals from Coal

Coal is the most abundant fossil fuel. Prior to the advent of petroleum as the major raw material for chemicals manufacture, coal tar was an important source of chemicals. Coal tar is a byproduct of coke manufacture for the steel industry. Thus, when coal is heated, in the absence of air, to about 1000°C coke is formed together with various gaseous and liquid decomposition products that include benzene, toluene, xylenes and naphthalene. The latter is still derived mainly from coal tar.

The last few years have witnessed a revival of interest in coal as a source of chemicals and liquid fuels. The economic viability of coal *vs.* oil-based fuels and chemicals depends on the oil-to-coal price ratio and this has increased considerably during the past few years and will probably continue to increase. Figure 1.4 illustrates predicted oil and coal production levels through the 23rd Century [10].

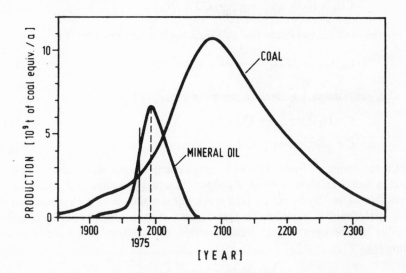

Figure 1.4. Predicted world production of coal and oil. Taken from H. Schulz, ref. [10], by permission of the International Union of Pure and Applied Chemistry.

There are two principal routes that have been used for converting coal to liquid hydrocarbons: hydrogenation (Bergius process) and gasification to syn gas followed by catalytic conversion to hydrocarbons (Fischer-Tropsch process). Both processes originated in Germany where they were operated on a large scale during World War II. The Fischer–Tropsch process is still used on a large scale in South Africa for the conversion of coal to gasoline and chemicals. In both instances the overriding factor in justifying these operations was strategic rather than economic. The Fischer-Tropsch process is discussed in more detail in Chapter 3 together with other routes for syn gas conversion to hydrocarbon feedstocks. Coal liquefaction via hydrogenation involves catalytic hydrogenation at high pressures in the presence of a solvent. A detailed discussion of this topic is beyond the scope of this book, where we shall be concerned with chemicals derived from syn gas.

1.5 Chemicals from Syn Gas

SYN GAS GENERATION

Three methods are used industrially for generating syn gas:

a. *Steam reforming of methane*

$$CH_4 + H_2O \xrightarrow[850°C/30\ bar]{Ni} CO + 3H_2 \tag{9}$$

b. *Partial oxidation of heavy fuel oil (Shell Gasification Process)* [11]

$$C_nH_{2n} + {}^n/_2\ O_2 \longrightarrow n\ CO + n\ H_2 \tag{10}$$

c. *Coal gasification (e.g. Shell–Koppers process)* [12]

$$C + H_2O \longrightarrow CO + H_2 \tag{11}$$

$$C + \tfrac{1}{2}O_2 \longrightarrow CO \tag{12}$$

Steam reforming of hydrocarbons involves reaction with steam over a hetero-geneous nickel catalyst at elevated temperatures and pressures. It is, in principle, applicable to any hydrocarbon feed stock but is mainly used with methane and, to a lesser extent, naphtha. In this context it is worth mentioning another possible source of syn gas: toluene steam dealkylation (i.e. toluene steam reforming):

$$PhCH_3 + H_2O \xrightarrow[430°C]{Rh/Al_2O_3} PhH + CO + 2\ H_2 \tag{13}$$

Similarly, the partial oxidation method can be applied to any hydrocarbon feedstock. In practice heavy fuel oil is generally used [11]. Coal gasification is actually a combination of steam reforming and partial oxidation (see Equations 11 and 12). The former is a strongly endothermic reaction and the latter is strongly exothermic. There are three well-proven processes for coal gasification: Shell–Koppers, Winkler and Lurgi. All three processes were developed in

Germany. The Shell–Koppers process is reported [12] to give practically complete gasification of virtually all solid fuels. A *ca.* 2 : 1 mixture of CO/H_2 is formed in 93–98% yield with little byproduct formation.

The different methods of syn gas generation produce various molar ratios of CO/H_2, summarized in Table 1.IV. This ratio can be important when considering

TABLE 1.IV
Syn gas ratios from different sources

Source	CO	H_2
1. Methane steam reforming	1	3
2. Naphtha steam reforming	1	2
3. Partial oxidation	1	1
4. Coal gasification	2	1

further conversion processes. For example, syn gas for methanol manufacture, which requires a CO/H_2 molar ratio of 1 : 2, is currently produced by methane steam reforming. Syn gas for hydroformylation, on the other hand, is produced by partial oxidation which affords the required 1 : 1 molar ratio. The CO/H_2 molar ratio can be adjusted to that required by use of the water gas shift reaction (WGSR):

$$CO + H_2O \rightleftharpoons CO_2 + H_2 \tag{14}$$

The reaction is carried out commercially [13] over supported metal oxide catalysts at elevated temperatures. The most common type is based on Fe_3O_4 promoted with Cr_2O_3, and it operates at about $350°C$. A copper/zinc oxide-based catalyst is more active but considerably more susceptible to sulphur poisoning. This has resulted in much attention being devoted [14] to the development of homogeneous transition metal catalysts for the water gas shift reaction (see Chapter 2 for a discussion of mechanism).

When an increase in the H_2/CO ratio is required this is accompanied by the formation of prodigious amounts of carbon dioxide at the expense of lost carbon monoxide. An alternative approach is to separate the CO and H_2 by selective absorption. The Cosorb process, for example, utilizes selective absorption of CO by a toluene solution of $CuAlCl_4$ [15]. Selective absorption has the advantage that no valuable CO is converted to unwanted CO_2.

Although steam reforming of methane and partial oxidation of heavy fuel oil presently account for the large part of syn gas production, coal gasification is expected to become an important source in the next five years. The second generation coal gasification processes mentioned above are much more efficient than the processes originally used in the 1930s. Coal gasification may be integrated with a combined-cycle power station, featuring both gas and steam turbines, for electricity generation. Some of the syn gas could then be utilized for the production of chemicals and liquid fuels. Schemes have also been proposed for underground gasification of less accessible coal deposits [16]. The production of syn gas by gasification is moreover, not restricted to coal. It can also be applied to other carbon sources such as biomass. Thus, syn gas may in the future be produced by gasification of urban or agricultural wastes or even mixtures of coal and biomass. There seems to be little doubt, therefore, that syn gas will become abundantly available in the near future.

EXISTING SYN GAS-BASED PROCESSES

Syn gas from natural gas steam reforming is used industrially for the manufacture of enormous quantities of methanol and ammonia:

$$CO + 2H_2 \;\rightleftharpoons\; CH_3OH \tag{15}$$

$$N_2 + 3H_2 \;\rightleftharpoons\; 2NH_3 \tag{16}$$

These in turn are converted to downstream derivatives such as formaldehyde and urea. These processes are usually integrated in a single chemical complex. They constitute a prime example of how the chemical industry, by process integration, maximizes the utilization of a feedstock (see Figure 1.5). In recent years the manufacture of acetic acid via carbonylation of methanol (Monsanto

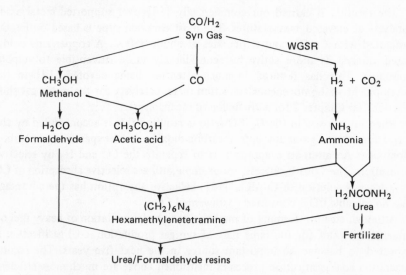

Figure 1.5.   Existing processes for chemicals from syn gas/methanol.

process; Reaction 17) has become increasingly important commercially. It is gradually displacing the traditional routes based on n-butane and ethylene oxidation (see Chapter 7 for a more detailed discussion).

$$CH_3OH + CO \xrightarrow{Rh^I/CH_3I} CH_3CO_2H \qquad (17)$$

Syn gas from partial oxidation ($CO/H_2 = 1/1$) is used for the manufacture of so-called OXO alcohols via the hydroformylation of appropriate olefins (Oxo process).

$$RCH=CH_2 + CO + H_2 \xrightarrow{catalyst} RCH_2CH_2CHO$$
$$\xrightarrow{H_2} RCH_2CH_2CH_2OH \qquad (18)$$

The hydroformylation reaction, which is of great importance industrially, is discussed in more detail in Chapter 4. In addition to the above mentioned uses carbon monoxide and hydrogen find industrial application in the production of a wide variety of organic chemicals via carbonylation and hydrogenation, respectively.

1.6 General Considerations for Future Syn Gas/Methanol-Based Chemicals Manufacture

There are basically two strategies for developing syn gas as a feedstock for chemicals. The first involves converting syn gas to ethylene and the other hydrocarbon building blocks mentioned earlier. These can be subsequently converted to industrial chemicals using the established technology outlined in Section 3. The second strategy involves the direct conversion of syn gas, or syn gas-derived methanol, to industrial chemicals without the intermediacy of ethylene, etc.

Several different pathways can be envisaged for the conversion of syn gas to ethylene. These are outlined in Figure 1.6. The conventional route, as operated in South Africa, is to convert the syn gas, via the Fischer–Tropsch process and subsequent distillation, to naphtha followed by steam cracking. Much research is also being devoted to the direct conversion of syn gas to light olefins (see Chapter 3). Alternatively, the syn gas can be converted to methanol by well established technology. The methanol can then be "cracked" to ethylene over zeolite catalysts (see Chapter 3) or converted to ethanol by reaction with more syn gas (homologation). Dehydration of the ethanol affords ethylene. For completeness it is also worth mentioning that ethanol (and methanol) may in the future be produced on a large scale from biomass, especially in countries like Brazil where the climate is suited to biomass production. The various routes to hydrocarbon fuels and chemicals are discussed in more detail in Chapter 3.

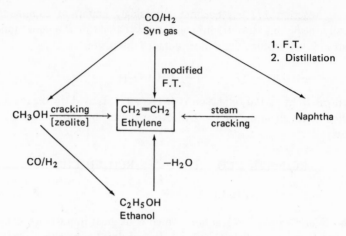

Figure 1.6. Alternative future sources of ethylene.

Two pathways can be delineated for the conversion of syn gas to industrial chemicals without going via ethylene. Firstly, by direct conversion as in the Union Carbide process for syn gas to ethylene glycol (see Chapter 9). Secondly, by conversion to methanol. The latter then serves as the basic building block for the manufacture of industrial chemicals. It is worth noting that methanol constitutes, in many respects, an ideal base chemical. It is for example, an easily transportable liquid that could be produced at a power plant and shipped to another location for further processing. Methanol may, in the future, also become important in its own right, as a gasoline extender or as a synthetic motor fuel (syn fuel) itself. As can be seen in Figure 1.7 methanol occupies an equivalent position to ethylene in the sequence : fossil fuel → feedstock → base chemical. An important consideration in determining a strategy for chemicals

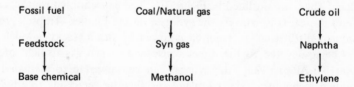

Figure 1.7. Methanol vs ethylene as a base chemical.

manufacture is the effectiveness of syn gas utilization. If we consider the syn gas utilization, on a weight basis, for the production of various industrial chemicals (see Table 1.V) it is readily apparent that direct conversion to oxygenated derivatives is much more effective than conversion to ethylene.

TABLE 1.V
Syn gas utilization in chemicals production

Product	H_2/CO Molar ratio	Wt. retention (%)
Methanol	$2 : 1$	100
Ethylene glycol	$1\frac{1}{2} : 1$	100
Acetic acid	$1 : 1$	100
Acetic anhydride	$1 : 1$	85
Ethyl acetate	$1\frac{1}{2} : 1$	71
Vinyl acetate	$1\frac{1}{4} : 1$	70
Ethanol	$2 : 1$	72
Ethylene	$2 : 1$	44
BTX	$1\frac{1}{2} : 1$	42

In the hypothetical production of ethylene, for example, 56% of the feedstock weight is lost as water:

$$2\,CO + 4\,H_2 \longrightarrow CH_2{=}CH_2 + 2\,H_2O \qquad (19)$$

Direct conversion to ethylene glycol or acetic acid, on the other hand, utilizes 100% of the syn gas feedstock:

$$2\,CO + 3\,H_2 \longrightarrow HOCH_2CH_2OH \qquad (20)$$

$$2\,CO + 2\,H_2 \longrightarrow CH_3CO_2H \qquad (21)$$

Products such as vinyl acetate and ethanol occupy an intermediate position:

$$4\,CO + 5\,H_2 \longrightarrow CH_2{=}CHOAc + H_2O \qquad (22)$$

$$2\,CO + 4\,H_2 \longrightarrow CH_3CH_2OH + H_2O \qquad (23)$$

Thus, we may conclude that direct conversion of syn gas/methanol to oxygenated derivatives appears to be a more economical route than one involving the intermediacy of ethylene. It should be noted, however, that in order to make a strictly valid comparison the energy consumptions of the various processes should also be included to give the total picture. For some products, e.g. polyethylene, it is difficult (but not impossible) to envisage non-ethylene based routes, which suggests that it is unlikely that ethylene will be completely eliminated as a base chemical.

1.7 The Role of Transition Metal Catalysis

In the event that syn gas achieves prominence as a basic raw material for chemicals manufacture, transition metal catalysis will have played an indispensable role in the development of the underlying chemistry. Virtually all of the reactions of carbon monoxide and hydrogen proceed only in the presence of appropriate transition metal catalysts that are capable of activating the inert, but intrinsically reactive, CO and H_2 molecules.

The early work on syn gas was generally concerned with conversion to hydrocarbons, e.g. via the Fischer–Tropsch process, and involved heterogeneous catalysis. More recent developments, on the other hand, are concentrated on the direct conversion of syn gas to oxygenated derivatives and invariably involve the use of soluble, i.e. homogeneous, transition metal complexes as catalysts. Indeed, the catalytic reactions of CO and H_2 have occupied a central position in the development of homogeneous transition metal catalysis over the last forty years. Thus, much of the chemistry discussed in this book has its origins in two basic discoveries made in Germany in the late 1930s. In 1938 Roelen of Ruhrchemie found that olefins can be converted to aldehydes by treatment with syn gas in the presence of a homogeneous cobalt catalyst at elevated temperatures and pressures [17].

$$RCH{=}CH_2 + CO + H_2 \longrightarrow RCH_2CH_2CHO \qquad (24)$$

This process, known as hydroformylation, is now one of the most important industrial applications of homogeneous transition metal catalysis (see Chapter 4). The second major breakthrough was the discovery, by Reppe of IG Farben, that soluble metal carbonyl complexes are able to catalyse a host of carbonylation reactions involving olefins, acetylenes or alcohols as substrates [18], e.g.

$$RCH{=}CH_2 + CO + H_2O \longrightarrow RCH_2CH_2CO_2H \qquad (25)$$

$$RCH_2OH + CO \longrightarrow RCH_2CO_2H \qquad (26)$$

Homogeneous, liquid-phase processes possess several advantages over their heterogeneous, gas-phase counterparts. Thus, homogeneous catalysts usually operate under milder conditions and are, consequently, more selective. Temperature and mixing are usually better controlled and catalyst and ligand concentrations are more easily regulated than in heterogeneous systems. At a time when society at large, and the chemical industry in particular, is much concerned with the conservation of energy and raw materials these superior activities and selectivities have become very relevant. Homogeneous catalysts are also more amenable to "fine tuning" by slight changes in the ligand environment around the metal. Being discrete chemical entities, homogeneous catalysts lend themselves readily to mechanistic studies using spectroscopic techniques.

The major advantage of heterogeneous catalysts, on the other hand, is the ease of separation and recovery of the catalyst from the products. One solution to this problem is to anchor a soluble catalyst to an insoluble support. Thus, when a soluble transition metal complex is attached to an insoluble support through covalent bonding the resulting "hybrid" catalyst is a heterogeneous catalyst when considered at the bulk level but is essentially identical to a homogeneous catalyst at the molecular level [19–21]. These catalysts can, in principle, combine the virtues of homogeneous catalysts – higher activity and selectivity – with the ease of separation characteristic of heterogeneous catalysts. The most commonly used supports are silica, alumina and cross-linked polystyrene resins modified with pendant phosphine ligands [19–21]. Much effort has been devoted, for example, to the development of hybrid rhodium catalysts for olefin hydroformylation [22]. A more recent example is the use of cobalt carbonyl complexes anchored to a cross-linked divinylbenzene–polystyrene resin as hybrid Fischer–Tropsch catalysts [23]. These hybrid catalysts will probably find wide industrial application in the future.

Another approach to the problem of catalyst separation is to carry out the reaction in a two-phase, liquid–liquid system in which the catalyst and products are dissolved in different phases. Phase transfer catalysis can then be used for shuttling the catalyst back and forth between phases [24]. This technique has been applied with success in carbonylation reactions and is described in more detail in Chapter 5.

1.8 Reaction Types

As noted earlier syn gas can be used as such in chemicals manufacture or as a source of pure hydrogen or carbon monoxide. The reactions of syn gas can, thus, be divided into four types on the basis of the reactants involved:

(a) Reaction of H_2 with a substrate (*hydrogenation*) e.g.

$$RCH{=}CH_2 + H_2 \longrightarrow RCH_2CH_3 \tag{27}$$

(b) Reactions of CO with a substrate (*carbonylation*) e.g.

$$CH_3OH + CO \longrightarrow CH_3CO_2H \tag{28}$$

(c) Reaction of CO and H_2 with a substrate (*hydroformylation* and *hydrocarbonylation*) e.g.

$$RCH{=}CH_2 + CO + H_2 \longrightarrow RCH_2CH_2CHO \tag{29}$$

$$CH_3OH + CO + 2\,H_2 \longrightarrow CH_3CH_2OH + H_2O \tag{30}$$

(d) Reaction of CO and H_2 (*CO hydrogenation*) e.g.

$$CO + 2\,H_2 \longrightarrow CH_3OH \tag{31}$$

$$n\,CO + 2n\,H_2 \longrightarrow C_nH_{2n} + n\,H_2O \tag{32}$$

A thorough treatment of hydrogenation (type a) is beyond the scope of this book and the reader is referred to the several standard texts on catalytic hydrogenation in homogeneous [25, 26] and heterogeneous [27, 28] systems. Hydrogenation is dealt with only in the context of syn gas chemistry and where it is relevant to mechanistic discussions.

In our treatment of syn gas chemistry we begin with a discussion of conversion to hydrocarbons. These reactions produce essentially the same building blocks as those that are currently used in the oil-based chemical industry. Such an approach obviously represents the least radical change in the present structure of the industry.

This is followed (Chapters 4 and 5) by a discussion of the various reactions of hydrocarbon building blocks with syn gas or carbon monoxide to give oxygen-containing derivatives. The remainder of the book deals with the direct conversion of syn gas or syn gas-derived methanol to oxygen-containing derivatives. This comprises some of the most novel and commercially interesting chemistry of syn gas, much of which has been discovered only in the last few years.

Before going on to discuss the reactions of syn gas in detail it is necessary first to provide a sound foundation for discussion by looking at the mechanisms of the fundamental steps involved. This we shall do in the next chapter.

References

1. L. F. Hatch and S. Matar, *From Hydrocarbons to Petrochemicals*, Gulf Publishing Co., Houston, 1981.
2. C. Ellis, *Chem. Met. Eng.*, **23**, 1230 (1920); see also C. Ellis, *Chemistry of Petroleum Derivatives*, Chemical Catalog Co., New York, 1934; B. T. Brooks, *Ind. Eng. Chem.*, **27**, 282 (1935).
3. P. Arthur and B. C. Pratt, *US Patent* 2,571,099 (1951); W.C. Drinkard and R. V. Lindsey, *US Patent* 3,496,215 (1970).
4. F. W. Thuener, *Chem. Econ. Eng. Rev.*, **11** (7), 23 (1979); A. Clementi, G. Oriani, F. Ancillotti and G. Pecci, *Hydrocarbon Process*, **58** (12), 109 (1979).
5. R. A. Sheldon, *J. Mol. Catal.*, **7**, 107 (1980); see also R. A. Sheldon, in *Aspects of Homogeneous Catalysis*, (R. Ugo, Ed.) Vol. 4, Reidel, Dordrecht, 1981, p. 3.
6. R. Landau, G. A. Sullivan and D. Brown, *Chem. Tech.*, 602 (1979).
7. A. P. Gelbein and A.S. Nislick, *Hydrocarbon Process.*, **57** (11) 125 (1978); see also W. W. Kaeding, *Hydrocarbon Process.*, **43** (11), 173 (1964).
8. R. A. Innes and H. E. Swift, *Chem. Tech.*, 244 (1981).
9. S. Fujiyama and T. Kasahara, *Hydrocarbon Process.*, **57** (11), 147 (1978).
10. H. Schulz, *Pure Appl. Chem.*, **51**, 2225 (1979).
11. C. L. Reed and C. J. Kuhre, *Hydrocarbon Process.*, **58** (9), 191 (1979).
12. H. K. Volkel. *Chem. Ind.*, **31**, 821 (1979); E. V. Vogt and M. J. van der Burgt, paper presented at the 72nd AIChE meeting, San Francisco, Cal., 29 Nov., 1979; H Staege, *Erdol Erdgas Zeit.*, **95**, 390 (1979).
13. C. L. Thomas *Catalytic Processes and Proven Catalysts*, Academic Press. New York, 1970, p. 104.

14. P. C. Ford, *Acc. Chem. Res.*, **14**, 31 (1981).
15. *Chem. Econ. Eng. Rev.*, 9 (12), 29 (1977); *Chem. Eng.*, Dec. 5, 1977, p. 122.
16. *C. and E. News*, July 21, 1980, p. 37; *C. and E. News*, Dec. 3, 1979, p. 19.
17. O. Roelen, *German Patent* 849,548 (1938) to Ruhrchemie; O. Roelen, *Angew. Chem.*, **60**, 62 (1948).
18. W. Reppe, *Justus Liebig's Ann. Chem.*, **582**, 1 (1953).
19. R. H. Grubbs, *Chem. Tech.*, 7, 512 (1977); R. H. Grubbs and S. C. H. Su, in *Enzymic and Non-Enzymic Catalysis*, (P. Dunnill, A. Wiseman and N. Blakebrough, Eds.) Horwood, Chichester, 1980. Chapter 9.
20. Z. M. Michalska and D. E. Webster, *Chem. Tech.*, 117 (1975).
21. D. C. Bailey and S. H. Langer, *Chem. Rev.*, **81**, 109 (1981).
22. K. G. Allum, R. D. Hancock, I. V. Howell, R. C. Pitkethly and P. J. Robinson, *J. Catal.*, **43**, 322 (1976).
23. K. P. C. Vollhardt and P. Perkins, *US Patent* Appl. 39,986 (1980) see *CA*, **94**, 128169v (1981).
24. H. Alper, *Advan. Organometal. Chem.*, **19**, 183 (1981).
25. B. R. James, *Homogeneous Hydrogenation*, Wiley, New York, 1973.
26. F. J. McQuillin, *Homogeneous Hydrogenation in Organic Chemistry*, Reidel, Dordrecht, 1976.
27. P. N. Rylander, *Catalytic Hydrogenation in Organic Synthesis*, Academic Press, New York, 1979.
28. M. Freifelder, *Catalytic Hydrogenation in Organic Synthesis – Procedures and Commentary*, Wiley-Interscience, New York, 1978.

ADDITIONAL READING

H. A. Witcoff and B. G. Reuben, *Industrial Organic Chemicals in Perspective*; Part. 1, *Raw Materials and Manufacture*, Wiley, New York, 1980.

A. M. Brownstein, *Trends in Petrochemical Technology*, Petroleum Publishing Co., Tulsa, Oklahoma, 1976.

K. Griesbaum and W. Swodenk, 'Progress and Trends in Petrochemistry', *Erdol Kohle Erdgas Petrochem.*, **33** (1), 34 (1980)

L. F. Hatch and S. Matar, *From Hydrocarbons to Petrochemicals*, Gulf Publishing Co., Houston, 1981.

H. Grunewald, Chemical Feedstocks – Problems of Today and Future Prospects', *Chem. Ind.* (London), p. 806, 1979.

C. D. Frohning and B. Cornils, 'Chemical Feedstocks from Coal, *Hydrocarbon Process.*, **53** (11), 143 (1974).

K. V. Rao and I. Skeist, 'Coal-derived Chemicals could open New Frontiers', *Oil Gas J.*, Feb. 1976, p. 90.

J. Gibson, 'Chemicals from Coal', *Chem, Brit.*, Jan. 1980, p. 26.

A. P. Gelbein, 'Perspective on Synthesis Gas Applications', *Process Development Digest*, 1 (3), 1 (1979) published by Chem Systems.

J. P. Leonard and M. E. Frank, 'The Coal-based Chemical Complex', *Chem. Eng. Progr.*, June 1979, p. 68.

J. Haggin, 'C_1 Chemistry Development Intensifies', *C. and E. News*, Feb. 23, 1981, p. 39.

J. Falbe, *Chemical Feedstocks from Coal*, Wiley, New York, 1981; *New Syntheses with Carbon Monoxide*, (J. Falbe, Ed.) Springer–Verlag, Berlin 1980.

I. Wender, 'Catalytic Synthesis of Chemicals from Coal', *Catal. Rev.*, **14**, 97 (1976).

Catalytic Activation of Carbon Monoxide, (P. C. Ford, Ed.) ACS Symposium Series, No. 152, 1981.

Catalysis of CO/H₂ Reactions. A Critical Analysis, Parts 1–5, Catalytica Associates Multiclient Study No. 1043, January 1978.

G. A. Mills, 'Catalytic Concepts in Coal Conversion', *Chem. Tech.*, 294 (1982).

M. E. Frank, 'Methanol: Emerging Uses, New Syntheses', *Chem. Tech.*, 358 (1982).

CHAPTER 2

MECHANISTIC PRINCIPLES

In this chapter we shall review, from a mechanistic viewpoint, the fundamental processes involved in reactions catalysed by transition metals. Particular emphasis is placed on those reactions pertaining to syn gas chemistry. More in-depth treatments of particular mechanisms are given in later chapters which deal with specific reactions of syn gas.

2.1 Characteristics of Catalytic Processes

The essential characteristics of a catalytic process are the same whether the catalyst is homogeneous or heterogeneous. Catalysis is a purely kinetic pheno-menon. A catalyst cannot make a thermodynamically forbidden reaction favour-able or alter the position of the final equilibrium. It can, however, dramatically accelerate a thermodynamically allowed, but kinetically slow, process by pro-viding an alternative, low-energy pathway for reaction. The catalyst lowers the free energy of activation of the rate-limiting step in the overall process.

The functions of a catalyst can be illustrated by reference to the potential energy surface for various reactions of CO and H_2 (see Figure 2.1). Combina-tions of stable compounds, such as hydrocarbons and water, correspond to deep wells in the energy surface. The states through which reactants pass correspond to troughs traversing the energy surface and connecting the deep wells belonging to the stable entities.

The height of the trough above the initial state corresponds to the activation energy of that particular reaction pathway. The width of the trough is related to the pre-exponential factor of the rate expression and the difference between the width of the trough and the width of the well corresponding to the initial state is related to the activation entropy.

Selectivity problems arise when the energy surface contains two or more final states, e.g. in Figure 2.1 the final states are CH_3OH, $CH_4 + H_2O$ and $C_nH_{2n+2} + H_2O$. A selective catalyst provides a deep trough only to the well that corresponds to the desired reaction products. In the absence of a catalyst the energy surface generally displays only shallow troughs connecting the wells. The activation energies are too high for the reaction to proceed at temperatures where the thermodynamic equilibrium is favourable.

21

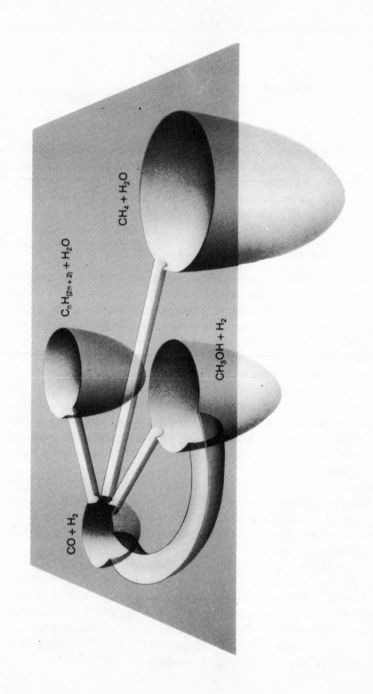

Figure 2.1. Energy Surface with Selective Catalyst. CO and H_2 can react to CH_4, CH_3OH and to higher hydrocarbons according to Fischer-Tropsch. A selective catalyst provides a path to one of the reaction products only (from *Harshaw Catalyst Resource Book*, Catalyst Dept., The Harshaw Chemical Co., Cleveland, Ohio, U.S.A.).

A catalyst provides a more favourable pathway (deeper trough) between the wells corresponding to the initial and final states of a chemical reaction. It does so by forming a chemical bond with an entity situated between the initial and final states. Chemical bonding decreases the energy of the transition state, i.e., it deepens the trough. A catalyst, therefore, increases the reaction rate constant, k, by decreasing the activation energy E in the expression:

$$k = k_0 \exp -E/RT$$

Since the decrease in activation energy is governed by the extent of bonding of reaction intermediates to the catalyst, a high bond strength would appear to be attractive. This is not the case, however. If a reaction intermediate bonds too strongly to the catalyst this may lead to a more stable state (deeper well) than that corresponding to the desired reaction products. In this case dissociation of the intermediate complex into reaction products becomes the rate-limiting step in the process. Thus, the strength of interaction of the catalyst with the reaction intermediates is critical, efficient catalysis being observed only within a fairly narrow range of favourable bonding energies. Moreover, since most of the reactions we shall be considering are multistep processes an efficient catalyst must be capable of effectively lowering the activation energy of all steps that do not occur readily in the absence of a catalyst.

The word catalyst is derived from Greek roots which mean "loosening". The Chinese word for catalyst is "Tsoo Mei" which means literally "matchmaker". These two apparently different concepts obviously refer to the bond-breaking and bond-making processes which form an integral part of all catalytic processes. The catalyst is, by definition, not consumed in the process but merely acts as a "matchmaker" in bringing the reactants together. It should be pointed out, however, that the active catalytic species is often not the same compound that is added to the reaction as a "catalyst". Thus, catalytic reactions often display an induction period during which the catalyst precursor is converted to the actual catalytic species.

Why are transition metals able to catalyse such a wide variety of reactions? Several important factors that contribute to their catalytic versatility are summarized below.

(a) The ability of transition metals to activate, via coordination, relatively inert molecules such as carbon monoxide.
(b) The ability of transition metals to stabilize a variety of unstable intermediates, e.g. as in metal hydrides and metal alkyls, in relatively stable but kinetically reactive complexes.
(c) The accessibility of different oxidation states and coordination numbers.

The propensity of transition metals to promote rearrangements, via ligand migration reactions within their coordination sphere, is dependent on these properties.

(d) The ability of transition metals to assemble and orient several reaction components within their coordination sphere (template effect). Syn gas reactions often involve the simultaneous coordination of CO, H_2 and a third molecule, such as an olefin, prior to reaction.

(e) The ability of transition metals to accommodate both participative and non-participative ligands. Modification of the steric and electronic properties of non-participative ligands (e.g. phosphines) can have a dramatic effect on catalytic activity and selectivity. It also allows for the "fine tuning" of the catalytic properties of transition metal complexes.

In general transition metal-catalysed processes can be viewed as occuring through a sequence of steps that incorporate the above features. These steps can be summarized as follows:

(a) Ligand dissociation/replacement
(b) Substrate binding and activation
(c) Ligand migration/insertion
(d) Product elimination

Each of these steps generally involves a change in the coordination number and/or oxidation state of the metal atom. Regeneration of the catalytically active species usually occurs in the product elimination step. The above sequence of reactions can be illustrated by taking as an example the homogeneous hydrogenation of an olefin in the presence of Wilkinson's catalyst, $RhCl(Ph_3P)_3$. This is known to involve the following sequence of steps [1].

Ligand dissociation $L_3Rh^ICl \rightleftharpoons L_2Rh^ICl + L$ (1)

H$_2$ activation $L_2Rh^ICl + H_2 \rightleftharpoons L_2(Cl)Rh^{III}\diagup^H_{\diagdown H}$ (2)

Olefin activation $L_2(Cl)Rh^{III}\diagup^H_{\diagdown H} + RCH=CH_2 \rightleftharpoons L_2(Cl)Rh^{III}\diagup^H_{\diagdown H}$ (3)

$$\underset{RCH=CH_2}{\uparrow}$$

Migratory insertion $L_2(Cl)Rh^{III}\diagup^H_{\diagdown H} \rightleftharpoons L_2(Cl)Rh^{III}\diagup^H_{\diagdown CH_2CH_2R}$ (4)

$$\underset{RCH=CH_2}{\uparrow}$$

Product dissociation

$$L_2(Cl)Rh^{III} \diagdown^{H}_{CH_2CH_2R} \longrightarrow L_2Rh^ICl + RCH_2CH_3 \qquad (5)$$

$$L = Ph_3P$$

Scheme 2.1

Ligand dissociation is followed by the oxidative addition (see below) of molecular hydrogen to form what is formally a rhodium(III) dihydride. Subsequent coordination of the olefin affords a rhodium-olefin π complex in which the double bond is activated towards nucleophilic attack by the suitably juxtaposed hydride ion. This leads to the formation of an alkylrhodium(III) complex by migratory insertion. The final step is then dissociation of the alkane product via reductive elimination (see below).

These fundamental steps will now be examined in more detail by taking examples that are pertinent to the catalytic reactions of carbon monoxide and hydrogen.

2.2 Metal-Ligand Interactions

BONDING IN TRANSITION METAL COMPLEXES

The d group transition elements possess nine valence shell orbitals — one s, three p and five d — that are potentially available for bonding to ligands through both covalent and coordinate bonds. The versatile chemistry of these elements is largely due to their propensity for forming both σ and π bonds to a variety of neutral molecules such as carbon monoxide, substituted phosphines and olefins. In the majority of these complexes the transition metal is in a low oxidation state, i.e. it is a "soft" metal centre according to Ugo's classification [2].

A characteristic feature of "soft", polarizable ligands, such as carbon monoxide and phosphines, is their ability to stabilize metals in low oxidation states. This property is associated with the fact that the donor atoms of these ligands possess vacant, low-lying orbitals in addition to lone pairs of electrons. Back donation of electrons from filled metal d orbitals to vacant, antibonding π^* orbitals on the ligand supplements the bonding arising from lone pair donation. This is illustrated for carbon monoxide binding in Figure 2.2.

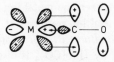

Figure 2.2. Interaction of metal and carbon monoxide orbitals.

High electron density on the low-valent metal can thus be delocalized onto the ligands. The ability of ligands to accept electron density into low-lying, vacant π orbitals is called π acidity and such ligands are called π-acids or π-acceptors. The most important π-acceptor ligand is undoubtedly carbon monoxide and carbonyl derivatives are known for all the transition elements. The most stable M—CO bonding results when a strong electron donor ligand, such as a substituted phosphine, is *trans* to the carbonyl ligand. In this configuration the R_3P and CO ligands share metal orbitals and electron density is transferred from R_3P to CO.

A similar type of synergistic bonding to that observed in metal carbonyls is also observed in the related metal-olefin complexes. Electron donation from the filled π orbitals of the olefin to a vacant, σ-type acceptor orbital on the metal is reinforced by back-bonding from filled π orbitals into antibonding π^* orbitals on the carbon atoms (see Figure 2.3). The overall polarity of the metal-carbon bond is dependent on the relative contributions of the two bonding

Figure 2.3. Interaction of metal and olefin orbitals.

modes. Thus, in complexes with metals in high oxidation states (2, 3 or higher) the olefin is susceptible to nucleophilic attack as a result of transfer of negative charge from the olefin to the metal. Metal-olefin complexes containing metals in low oxidation states (0, 1), on the other hand, are susceptible to electrophilic attack at carbon due to the overriding importance of back-bonding.

COORDINATIVE UNSATURATION

An important property that determines the catalytic behaviour of transition metal complexes is the presence of coordinative unsaturation, i.e. the presence of vacant coordination sites analogous to the "active sites" in heterogeneous systems. When the nine valence shell orbitals of the transition metal ion are fully utilized in bond formation, i.e. the total number of electrons involved in metal-ligand bonding is eighteen, the metal ion is said to be saturated. If the valence electron count is seventeen the metal ion is *covalently unsaturated* and if it is sixteen or less the metal ion possesses at least one vacant coordination site and is *coordinatively unsaturated*. For example, the first step in olefin hydroformylations catalysed by $HCo(CO)_4$ is dissociation of one CO ligand to form a reactive, coordinatively unsaturated species.

$$\text{HCo(CO)}_4 \;\rightleftharpoons\; \text{HCo(CO)}_3 + \text{CO} \qquad\qquad (6)$$

The facile formation of coordinatively unsaturated species is a characteristic feature of metals possessing many d electrons and containing a high electron density on the metal, i.e. metals in low oxidation states. This phenomenon has its origin in the fact that the higher the coordination number the fewer the d electrons which can be accomodated in stable, bonding molecular orbitals. In an octahedral complex, for example, the three non-bonding t_{2g} orbitals can accommodate up to $6d$ electrons (see Figure 2.4). Any additional electrons are forced to occupy the strongly antibonding $e_g{}^*$ orbitals. This explains why an electron configuration of more than $6d$ electrons leads to the destabilization of octahedral coordination in favour of a lower coordination number that permits the d electrons to be accommodated in stable orbitals.

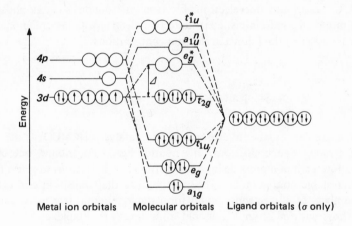

Figure 2.4. Molecular orbitals in an octahedral complex.

It is no coincidence, therefore, that the catalytic chemistry of CO and H_2 is dominated by the low-valent d^8 complexes of the iron, cobalt and nickel triads, in particular the first and second row elements, Fe, Ru, Co, Rh, Ni and Pd (see Figure 2.5). The d^8 configuration encompasses a wide range of stereo-chemical behaviour, coordination numbers of 4 and 5 being common. This

| Fe^0 | Co^I | Ni^{II} |
Ru^0	Rh^I	Pd^{II}
Os^0	Ir^I	Pt^{II}

Figure 2.5. d^8 Metal complexes.

allows for the facile formation of coordinatively unsaturated species and is a key factor in determining the catalytic activity of d^8 complexes.

THE 16 AND 18 ELECTRON RULE

The stability of coordination complexes of the transition elements can be predicted on the basis of the "18 electron rule" or "inert gas rule". The rule was originally applied to metal carbonyls but is broadly applicable to predicting the stability of organo-transition metal compounds. The rule predicts that a compound should be stable when the sum of the metal valence electrons and those contributed by the ligands is 18. This results in a stable, inert-gas configuration and the metal ion is saturated (see above).

Simple, covalently-bonded ligands such as halide, alkyl or hydride are considered to contribute one electron. Electron pair donors such as phosphines, carbon monoxide and olefins, contribute two electrons. Other common organic ligands are assigned the following electron contributions:

3	π-allyl
4	dienes
5	π-cyclopentadienyl
6	π-arene

For example, in the hydroformylation catalyst, $HCo(CO)_4$, the cobalt possesses nine valence electrons. The hydride ligand contributes one electron and the four carbon monoxide ligands contribute two electrons to give a total of 18, i.e. a stable configuration. The rule predicts that transition elements with even numbers of d electrons should form stable mononuclear metal carbonyls whilst those with odd numbers should form dinuclear complexes as is generally observed in practice, e.g. $Fe(CO)_5$, $Co_2(CO)_8$ and $Ni(CO)_4$.

The generality of the 18 electron rule for predicting stable configurations led to the formulation by Tolman [3] of the "16 and 18 electron rule" for identifying preferred reaction pathways in homogeneous catalysis. Basically the 16 and 18 electron rule states that a catalytic process will proceed via a sequence of ligand dissociation and association steps in which intermediates containing 16 and 18 valence electrons alternate. Application of the rule can be illustrated by reference to the generally accepted mechanism for the cobalt-catalysed hydroformylation of olefins which involves the following reaction sequence:

$$HCo^I(CO)_4 \xrightarrow{-CO} HCo^I(CO)_3 \xrightarrow{RCH=CH_2} HCo^I(CO)_3(RCH=CH_2)$$
$$\quad 18e \qquad\qquad\qquad 16e \qquad\qquad\qquad\qquad 18e$$

$$\longrightarrow RCH_2CH_2Co^I(CO)_3 \xrightarrow{CO} RCH_2CH_2Co^I(CO)_4$$
$$\qquad\qquad 16e \qquad\qquad\qquad\qquad 18e$$

$$\longrightarrow RCH_2CH_2COCo^I(CO)_3 \xrightarrow{H_2} RCH_2CH_2COCo^{III}(H_2)(CO)_3$$
$$16e \qquad\qquad\qquad 18e$$

$$\longrightarrow RCH_2CH_2CHO + HCo^I(CO)_3$$
$$16e$$

Scheme 2.2

Despite the broad utility of the 16 and 18 electron rule for predicting reaction pathways there are many exceptions to the rule. One conspicuous example is Wilkinson's catalyst, $RhCl(Ph_3P)_3$. As was mentioned earlier the first step in reactions of this 16 electron complex is dissociation of a triphenylphosphine ligand to give a highly reactive 14 electron complex (Reaction 1). In this and analogous d^8 square planar complexes dissociation to a 14 electron complex is promoted by the steric bulk of the triphenylphosphine ligands. It should also be noted that the 16 and 18 electron rule precludes the existence of reactive, paramagnetic intermediates in catalytic reactions. In the last few years, however, it has become increasingly apparent that many organometallic reactions proceed via discreet, one-equivalent changes and involve free radicals as intermediates [4, 5].

LIGAND SUBSTITUTION

As we have discussed above the first step in a catalytic cycle is generally substitution of an extant ligand by the substrate. The 16 and 18 electron rule predicts that this process should occur via initial ligand dissociation to a 16 electron intermediate. This process bears a formal resemblance to the S_N1 mechanism for nucleophilic substitution in carbon chemistry. A simple example of such a ligand substitution is the reaction of $Cr(CO)_6$ with triphenylphosphine which leads to the replacement of two CO ligands [6].

$$Cr(CO)_6 \xrightarrow[slow]{-CO} Cr(CO)_5 \xrightarrow{Ph_3P} Cr(CO)_5(Ph_3P)$$
$$18e \qquad\qquad 16e \qquad\qquad 18e$$

$$\xrightarrow{-CO} Cr(CO)_4(Ph_3P) \xrightarrow{Ph_3P} Cr(CO)_4(Ph_3P)_2$$
$$16e \qquad\qquad\qquad 18e$$

Scheme 2.3

The $Cr(CO)_6$ molecule is coordinatively saturated and, hence, resistant to

$S_N 2$ nucleophilic attack. Ligand substitution proceeds via rate-limiting dissociation to a reactive, 16 electron species. Such CO dissociation processes generally have high activation energies, temperatures in excess of $100°C$ being needed for a reasonable rate of reaction. The reaction can, however, be effected photochemically at ambient temperature [7].

Analogous substitution reactions of $Mo(CO)_6$ and $W(CO)_6$, in contrast, exhibit second-order kinetics and appear to involve an $S_N 2$ type displacement through a nominally seven coordinate intermediate [6].

Recently it has been shown that ligand substitutions in metal carbonyl complexes can also involve one-equivalent changes and paramagnetic, 17 electron intermediates. This mechanism appears to operate, for example, in ligand substitution processes in binuclear carbonyl complexes such as $Mn_2(CO)_{10}$, $Co_2(CO)_8$ and $Re_2(CO)_{10}$ [8, 9]. Thus, ligand substitution in the important catalyst precursor $Co_2(CO)_8$ involves rate-limiting, homolytic dissociation to the paramagnetic, 17 electron species, $Co(CO)_4$, as shown below [8].

$$Co_2(CO)_8 \underset{\longleftarrow}{\overset{h\nu \text{ or } \triangle}{\longrightarrow}} 2\ Co(CO)_4 \tag{7}$$

$$Co(CO)_4 + R_3P \longrightarrow R_3P\,Co(CO)_3 + CO \tag{8}$$

$$R_3PCo(CO)_3 + Co(CO)_4 \longrightarrow (R_3P)\,Co_2(CO)_7 \tag{9}$$

The reactive $Co(CO)_4$ intermediate is analogous to free radical intermediates observed in organic reactions and Reaction 7 can be compared to the homolytic dissociation of typical free radical initiators such as organic peroxides. Similarly, ligand substitution in $HRe(CO)_5$ is induced by light and retarded by air and typical radical scavengers such as hydroquinone [10]. These observations are consistent with a radical chain mechanism involving the paramagnetic $Re(CO)_5$ as the chain transfer agent in the propagation sequence:

$$Re(CO)_5 + L \longrightarrow Re(CO)_4 L + CO \tag{10}$$

$$Re(CO)_4 L + HRe(CO)_5 \longrightarrow HRe(CO)_4 L + Re(CO)_5, \text{ etc.} \tag{11}$$

Many more examples of homolytic mechanisms in metal carbonyl and related chemistry have been discussed by Kochi [5]. As more systems are subjected to mechanistic scrutiny it is becoming increasingly apparent that the role of one equivalent changes and paramagnetic intermediates in organometallic chemistry has been grossly underemphasized in the past. The significance of homolytic mechanisms in such ubiquitous reactions as oxidative addition and ligand insertion (see below) is just beginning to be appreciated.

Now that we have discussed the introduction of the substrate to the metal catalyst we can consider in more detail the activation of the substrate molecule.

2.3 Molecule Activation

Molecule activation is simply the process of rendering a substrate more reactive than it would otherwise be in the absence of the catalyst [11]. Coordination to a metal ion induces changes in the electron distribution of a ligand, such as carbon monoxide and olefins, that can dramatically modify its reactivity. As mentioned earlier this usually means transferring electron density from substrate bonding orbitals to the metal centre and back-donating electron density from the metal into antibonding orbitals on the substrate. For an X–Y substrate this can lead to cleavage of the X–Y bond in a process called *oxidative addition* (see later). When the XY bond is multiple in character substrate activation usually leads to a reduction in the X–Y bond order.

Ligand coordination not only influences the properties of the ligand but also those of the metal ion. The reactivity of the metal centre is influenced by both the electronic and steric characteristics of the surrounding ligands. Indeed many of the reactions of metal carbonyls involve attack of another substrate at the metal centre and not at the CO ligand, i.e. this can better be viewed as activation of the metal ion by CO rather than *vice versa*.

ACTIVATION OF CARBON MONOXIDE

Carbon monoxide is an extremely weak sigma donor ligand ("hard base"). It is, however, a good "soft" ligand and bonds quite tenaciously to transition metals in low oxidation states (see above for a discussion of the bonding). There are three primary modes of carbon monoxide coordination: terminal, bridging and triply bridging. These are illustrated in the valence bond structures below.

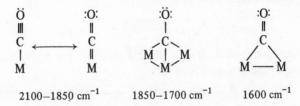

| IR stretching frequency | $2100-1850$ cm^{-1} | $1850-1700$ cm^{-1} | 1600 cm^{-1} |

The terminal mode of coordination is by far the most common one. Synergistic σ donation and π-back bonding between metal and ligand results in a reduction of the stretching frequency of the C–O bond from the 2155 cm^{-1} of free CO to the range 2100–1850 cm^{-1}. This lowering of the carbonyl stretching frequency reflects a reduction in the C–O bond order.

The bridging mode of coordination is often referred to as the "ketonic" mode because of the formal resemblance to organic carbonyl compounds. The CO group forms a covalent bond to two metal centres and the CO bond order is reduced to two, or less with additional back bonding, and the CO stretching frequency is in the range 1850–1700 cm^{-1}. The triply bridging mode of CO bonding is relatively rare. While a simple valence bond structure is grossly inadequate to describe the bonding in such complexes it is readily apparent that it results in a significant reduction in the CO bond order. The ketonic and triply bridging bonding modes can be compared to the coordination of CO to metal surfaces which can involve bonding to two or more adjacent metal atoms.

Coordination of CO to transition metal ions produces an electronic perturbation in the CO the extent of which is dependent on many factors such as the nature of the metal, its charge, oxidation state and the nature of other co-ordinated ligands. This perturbation generally leads to an increased reactivity towards attack by nucleophiles such as hydroxide, alkoxide and amines. The reaction of metal carbonyls with hydroxide ion, for example, is well known and provides a useful route to metal carbonyl hydrides. Thus, if iron pentacarbonyl is treated with aqueous sodium hydroxide it dissolves to give a solution of the salt, $Na^+[HFe(CO)_4]^-$, which on acidification affords a thermally unstable gas, $H_2Fe(CO)_4$ [12].

$$Fe(CO)_5 + HO^- \longrightarrow [(CO)_4Fe-CO_2H]^-$$

$$\xrightarrow{-CO_2} [HFe(CO)_4]^- \xrightarrow{H^+} H_2Fe(CO)_4$$

Scheme 2.5

This reaction is general for a variety of metal carbonyls and forms the basis for the homogeneous catalysis of the water gas shift reaction by metal carbonyls in basic media [13–19].

$$CO + H_2O \xrightarrow{[M(CO)_n]} CO_2 + H_2 \tag{12}$$

Catalysis by $Fe(CO)_5$ in alkaline medium, for example, proceeds via the following cyclic process [16, 17]:

$$Fe(CO)_5 + HO^- \longrightarrow HFe(CO)_4^- + CO_2 \tag{13}$$

$$HFe(CO)_4^- + H_2O \longrightarrow H_2Fe(CO)_4 + HO^- \tag{14}$$

$$H_2Fe(CO)_4 \longrightarrow Fe(CO)_4 + H_2 \tag{15}$$

$$Fe(CO)_4 + CO \longrightarrow Fe(CO)_5 \tag{16}$$

Scheme 2.6

When the metal is in a high oxidation state, transfer of electron density from CO

to the metal ion renders the former susceptible to attack, even by weak nucleophiles such as water. It is well known that CO in the presence of water is an
effective reducing agent for transition metal ions. In the case of rhodium(III),
James and coworkers [20] showed that the reaction involves nucleophilic attack
of water on coordinated CO to give an unstable hydroxycarbonylrhodium(III)
intermediate that decomposes to CO_2 and rhodium(I).

$$Rh^{III}CO + H_2O \longrightarrow Rh^{III}-\overset{\overset{\displaystyle O}{\|}}{C}-OH \longrightarrow Rh^{I} + CO_2 + H^+ \qquad (17)$$

Reaction (17) forms the basis for the rhodium-catalysed water gas shift
reaction in neutral and acidic media [21–23]. Analogous attack of an alcohol
on a palladium(II) carbonyl complex is the key step in the palladium-catalysed
oxidative carbonylation of alcohols (see Chapter 7).

$$XPd^{II}CO + ROH \longrightarrow Pd^{II}-O-\overset{\overset{\displaystyle O}{\|}}{C}R + HX \qquad (18)$$

Susceptibility to attack by nucleophiles is not restricted to external nucleophiles as in the examples above. Probably the most important modification of
CO reactivity is with respect to intramolecular nucleophilic attack by other
ligands, such as alkyl or hydride, coordinated to the same metal centre. This
leads to *migratory insertion* which is a key feature of many catalytic reactions of
CO (see Section 2.5).

The binding of CO to two or three metal centres can reduce the CO bond
order to such an extent that stable complexes can be formed between the
oxygen atom and Lewis acids such as trialkylaluminium compounds [24–27].

In a few instances terminal carbonyl ligands have been observed [28, 29] to
bind R_3Al with concomitant large reductions in the CO stretching frequency.
Activation of CO by simultaneous coordination of the carbon atom to a transition metal and the oxygen atom to Lewis acids may be relevant to catalytic
reactions of syn gas over transition metals on acidic supports.

Chemisorption of CO on metal surfaces often leads to metal carbide formation via carbon–oxygen bond cleavage (dissociative adsorption). With some
metals e.g. Mo, W, Ta, this process takes place even at very low temperatures
[30]. A prerequisite for C–O bond cleavage appears to be simultaneous co-

ordination to at least two metal centres, a condition readily satisfied on metal surfaces. Carbide formation may be the initiation step in the reductive oligomerization of CO by the Fischer–Tropsch process (see Section 2.6). It has also been observed in polynuclear metal carbonyls. Thus, heating polynuclear carbonyls such as $Ru_3(CO)_{12}$, $Os_3(CO)_{12}$ and $Fe_3(CO)_{12}$ to 200–300°C results in the formation of polynuclear carbide structures, such as $Ru_6C(CO)_{17}$ and $Fe_5C(CO)_{15}$, in which the carbon atom is located at the centre of a metal cage [31]. It was established that the carbide atom originates from a molecule of CO.

The activation of CO, although important for subsequent reactions, is not sufficient to produce a catalytic reaction. For this the simultaneous interaction of the catalyst with other molecules is needed. In the context of syn gas chemistry the other molecule of interest is hydrogen.

ACTIVATION OF HYDROGEN

The ability to activate molecular hydrogen has been demonstrated for a wide variety of soluble transition metal complexes. It also occurs readily on many metal surfaces. In each case activation leads to cleavage of the hydrogen molecule to give hydrido transition metal complexes. Three distinct mechanisms have generally been proposed to account for hydrogen activation by transition metal complexes [32, 33]. These are illustrated in Equations 19–21.

Oxidative Addition

$$Rh^ICl(Ph_3P)_2 + H_2 \longrightarrow Rh^{III}H_2(Cl)(Ph_3P)_2 \tag{19}$$

Heterolytic Addition

$$Ru^{II}Cl_2(Ph_3P)_2 + H_2 \longrightarrow Ru^{II}H(Cl)(Ph_3P)_2 + HCl \tag{20}$$

Homolytic Addition

$$2\,Co^{II}(CN)_5^{3-} + H_2 \longrightarrow 2\,Co^{III}(CN)_5H^{3-} \tag{21}$$

However, a closer inspection reveals that oxidative addition and heterolytic addition are in many, if not all, cases essentially the same mechanism. Thus, reaction (20) is promoted by bases and almost certainly involves initial oxidative addition followed by reductive elimination, i.e.,

$$Ru^{II}Cl_2L_2 + H_2 \longrightarrow Ru^{IV}H_2Cl_2L_2 \longrightarrow Ru^{II}HClL_2 + HCl \tag{22}$$

An important, fundamental question which still remains to be answered is whether or not hydrogen activation involves the same initial step in all of these processes, namely homolytic cleavage of the H–H bond. Whether this leads to overall oxidative or homolytic addition would then depend on the fate of the initially formed radical pair as illustrated below.

$$M^{n+} + H_2 \longrightarrow [H-M^{(n+1)+} + H\cdot] \begin{cases} \longrightarrow H_2 M^{(n+2)+} & (23) \\ \xrightarrow{M^{n+}} 2 H-M^{(n+1)+} & (24) \end{cases}$$

This is analogous to the corresponding reaction of transition metal complexes with molecular oxygen which, depending on the metal, leads to the formation of side-on (dihapto) peroxo or μ-peroxo complexes [34].

$$M^{n+} + O_2 \longrightarrow M^{(n+1)+}-O-O\cdot \begin{cases} \longrightarrow \overset{O}{\underset{O}{|}} M^{(n+2)+} & (25) \\ \longrightarrow M^{(n+1)+}-O-O-M^{(n+1)+} & (26) \end{cases}$$

Of direct relevance to the chemistry of syn gas is the activation of hydrogen by metal carbonyls. Activation generally occurs by means of initial thermal or photochemical [7] dissociation to produce coordinatively unsaturated species. Subsequent oxidative addition of hydrogen affords metal carbonyl hydrides. Dinuclear metal carbonyls react with molecular hydrogen by initial metal–metal bond cleavage to paramagnetic intermediates. For example, thermal [35] or photochemical [36] activation of H_2 by $Mn_2(CO)_{10}$ affords the manganese carbonyl hydride, $HMn(CO)_5$ via the following reaction sequence:

$$Mn_2(CO)_{10} \underset{}{\overset{h\nu \text{ or } \triangle}{\rightleftharpoons}} 2 \ Mn(CO)_5 \tag{27}$$

$$Mn(CO)_5 \rightleftharpoons Mn(CO)_4 \tag{28}$$

$$Mn(CO)_4 + H_2 \longrightarrow H_2 Mn(CO)_4 \tag{29}$$

$$H_2 Mn(CO)_4 \cdot + Mn(CO)_5 \longrightarrow HMn(CO)_5 + HMn(CO)_4 \text{ , etc.} \tag{30}$$

A similar mechanism was proposed for the photochemical activation of H_2 by $Re_2(CO)_{10}$ [36, 37]. An alternative pathway was proposed for the activation of H_2 by the hydroformylation catalyst precursor, $Co_2(CO)_8$ [38, 39]:

$$Co_2(CO)_8 \rightleftharpoons Co_2(CO)_7 + CO \tag{31}$$

$$Co_2(CO)_7 + H_2 \rightleftharpoons HCo(CO)_4 + HCo(CO)_3 \tag{32}$$

$$HCo(CO)_3 + CO \rightleftharpoons HCo(CO)_4 \tag{33}$$

Scheme 2.7.

METAL CARBONYL HYDRIDES

The metal carbonyl hydrides play a central role in the catalytic chemistry of syn gas. They are generally obtained by acidification of the corresponding alkali metal carbonylates (see Scheme 2.8) or by reaction of hydrogen with metal carbonyls as described above. Cobalt carbonyl hydride, $HCo(CO)_4$, can also be obtained from the direct reaction of cobalt compounds with carbon monoxide and hydrogen at elevated temperatures and pressures. The three methods of preparation of $HCo(CO)_4$ are illustrated in Scheme 2.8.

$$Na\,Co(CO)_4 \;+\; H^+$$
$$Co_2(CO)_8 \;+\; H_2 \qquad\qquad\longrightarrow\; HCo(CO)_4$$
$$Co \;+\; 4\,CO \;+\; \tfrac{1}{2}\,H_2 \quad \underset{150°C}{\overset{50\ bar}{}}$$

Scheme 2.8.

The metal carbonyl hydrides are thermally unstable in the absence of high pressures of carbon monoxide. $H_2Fe(CO)_4$ and $HCo(CO)_4$, for example, decompose readily above -10 and $-20°C$, respectively. They are also extremely reactive substances due to the labile nature of the M–H bond, i.e. they make ideal catalysts. They undergo facile air oxidation, for example, suggesting that the M–H bond readily undergoes homolytic cleavage. In aqueous solution they behave as acids, ionizing to the corresponding carbonylate anions (see Table 2.I). $HCo(CO)_4$, for example, is a very strong acid, a 0.2M aqueous solution being completely dissociated. Titration studies indicated that in methanol solution $HCo(CO)_4$ is as strong an acid as hydrochloric and nitric acid [40].

$HCo(CO)_4$ is thus sufficiently acidic to protonate methanol and the cobalt

TABLE 2.I
Acid dissociation constants of metal carbonyl hydrides

Compound	pK_a
$HMn(CO)_5$	6.1
$H_2Fe(CO)_4$	4.4 (K_1)
	13.4 (K_2)
$H\,Co(CO)_4$	ca. 0
$H\,Co(CO)_3Ph_3P$	6.9
CH_3CO_2H	4.8
Cl_3CCO_2H	0.65

Data taken from D.F. Shriver, *Acc. Chem. Res.*, **3**, 231 (1970).

carbonyl-catalysed homologation (hydrocarbonylation) of methanol [41] involves the formation of a methyl cobalt intermediate via protonation of methanol and subsequent S_N2 displacement of water by the cobalt carbonylate anion (Reaction 34). This is analogous to the formation of methyl bromide from methanol and hydrogen bromide.

$$CH_3\overset{+}{O}H_2 + Co(CO)_4^- \longrightarrow CH_3Co(CO)_4 + H_2O \qquad (34)$$

$$\text{c.f. } CH_3OH + HBr \longrightarrow CH_3Br + H_2O \qquad (35)$$

It is worth noting in this context that the usual formalism of regarding the polarization of metal hydride bonds as $M^+ H^-$ is certainly very arbitrary in the case of the metal carbonyl hydrides. As mentioned above the M–H bond in carbonyl hydrides is also susceptible to homolytic cleavage. The facile reduction of organic halides by carbonyl hydrides, for example, appears to involve homolytic hydrogen transfer. Thus, $HMn(CO)_4L$ (L = phosphines and phosphites) reacts spontaneously with a variety of alkyl chlorides and bromides [42].

$$HMn(CO)_4L + RX \longrightarrow RH + XMn(CO)_4L \qquad (36)$$

The products obtained with various substrates were strongly indicative of alkyl radical intermediates and the following free radical chain mechanism was proposed to account for the results.

$$\text{Initiation} \qquad HMn + RX \longrightarrow Mn + R\cdot + HX \qquad (37)$$

$$\text{Propagation} \qquad Mn + RX \longrightarrow MnX + R\cdot \qquad (38)$$

$$R\cdot + HMn \longrightarrow RH + Mn \qquad (39)$$

There are well-established precedents for such a propagation sequence as expressed in Equations 38 and 39. The free radical chain mechanism for the analogous reduction of alkyl halides by tin hydrides, for example, is well-known [43].

$$RX + R'_3Sn\cdot \longrightarrow R'_3SnX + R\cdot \qquad (40)$$

$$R\cdot + R_3SnH \longrightarrow RH + R'_3Sn\cdot , \text{ etc.} \qquad (41)$$

The involvement of free radical or ionic intermediates in the reactions of metal carbonyl hydrides may be dependent on a number of factors such as the nature of the substrate, solvent polarity, the presence of other coordinated ligands, etc. This point is discussed further in Section 2.5 in connection with the important reaction of carbonyl hydrides with olefins.

CARBONYLMETALLATE ANIONS

We have mentioned earlier (Section 2.3) that the reaction of metal carbonyls with bases such as alkoxide, hydroxide and amine, affords the corresponding carbonyl metallate anion, the conjugate base of the metal carbonyl hydride. Both mononuclear, e.g. $Fe(CO)_4{}^{2-}$, and polynuclear carbonylmetallate anions, e.g. $Fe_2(CO)_8{}^{2-}$, $Fe_3(CO)_{11}{}^{2-}$ and $Ru_3(CO)_{11}{}^{2-}$, are known as well as derivatives in which one or more of the CO groups is replaced by other ligands.

The carbonylmetallate anions are powerful nucleophiles, $Fe(CO)_4{}^{2-}$ for example, is comparable in nucleophilic character to PhS^- [5]. Consequently they readily undergo S_N2 substitution reactions with alkyl halides, e.g.

$$Fe(CO)_4{}^{2-} + RX \longrightarrow RFe(CO)_4{}^- + X^- \qquad (42)$$

The alkyltetracarbonylferrate ion formed in Reaction 42 undergoes a variety of synthetically useful transformations as shown in Scheme 2.9 [44].

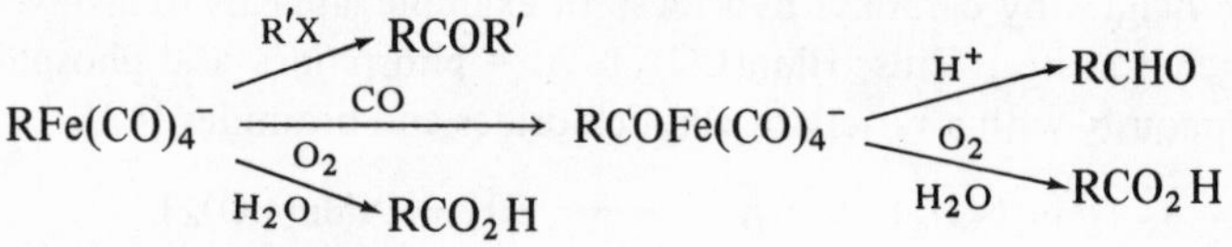

$$RFe(CO)_4{}^- \xrightarrow[\substack{O_2 \\ H_2O}]{\substack{R'X \\ CO}} \begin{array}{c} RCOR' \\ \\ RCO_2H \end{array} \quad RCOFe(CO)_4{}^- \xrightarrow[\substack{O_2 \\ H_2O}]{\substack{H^+ \\ }} \begin{array}{c} RCHO \\ \\ RCO_2H \end{array}$$

Scheme 2.9.

Carbonylmetallate anions may also be active intermediates in the catalytic hydrogenation of carbon monoxide in the presence of metal carbonyls as many of these reactions are carried out under basic conditions (see Section 2.6).

2.4. Oxidative Addition and Reductive Elimination

Oxidative addition is the generic name used to describe a class of reactions in which the oxidation of a metal complex by an electrophile is accompanied by an increase in its coordination number [5]. It is usually applied, without mechanistic implications, to reactions involving an overall two-equivalent change at the metal centre, and is described by the general equation,

$$\diagup M \diagdown + X{-}Y \longrightarrow \diagup M \diagdown_Y^X \quad \text{or} \quad \diagup M \diagdown_Y^X \qquad (43)$$

where X–Y can be H_2, O_2, hydrogen halide, halogens, organic halides, etc. The microscopic reverse process is known as *reductive elimination*.

Oxidative addition and reductive elimination are ubiquitous processes in organometallic chemistry and several review articles have been published on the subject [45–48]. Oxidative addition is a characteristic reaction of low-valent transition metal complexes having d^8 or d^{10} configurations. Two reactions of particular relevance to syn gas chemistry are the oxidative addition of H_2 (see previous section) and organic halides. The latter constitutes a synthetically useful method of forming alkylmetal complexes and plays an important role in a wide variety of catalytic processes. For example, the oxidative addition of methyl iodide to rhodium(I) is the key step in the iodide-promoted carbonylation of methanol in the presence of rhodium catalysts (see Chapter 7).

$$CH_3I + Rh^I \longrightarrow CH_3-Rh^{III}-I \qquad (44)$$

Reaction (44) is followed by carbon monoxide insertion (see following section) and reductive elimination of acetyl iodide which is hydrolysed to acetic acid (see Scheme 2.10).

$$CH_3Rh^{III}I + CO \longrightarrow CH_3CORh^{III}I \qquad (45)$$

$$CH_3CORh^{III}I \longrightarrow CH_3COI + Rh^I \qquad (46)$$

$$CH_3COI + H_2O \longrightarrow CH_3CO_2H + HI \qquad (47)$$

$$HI + CH_3OH \longrightarrow CH_3I + H_2O \qquad (48)$$

overall reaction $\quad CH_3OH + CO \xrightarrow{[HI,Rh^I]} CH_3CO_2H \qquad (49)$

Scheme 2.10

A number of factors determine the ease with which transition metal complexes undergo oxidative addition. The transition metal should be in a low oxidation state and behaves as a nucleophile or a reductant, oxidative addition resulting in a removal of electrons from the electron-rich metal centre. The reactivity of Group VIII metal complexes towards oxidative addition generally increases in going from right to left in the Periodic Table and in going down a particular triad. Thus, assuming a constant ligand environment the following order of reactivity is generally observed.

$$Os^0 > Ru^0 > Fe^0, \quad Ir^I > Rh^I > Co^I, \quad Pt^{II} > Pd^{II} > Ni^{II}$$

This order of reactivity parallels the ease with which the complexes undergo ligand dissociation to generate coordinatively unsaturated species. Indeed, prior dissociation to coordinatively unsaturated intermediates is often a requirement for oxidative addition reactions. The square-planar d^{10} complexes, $M(Ph_3P)_4$ (M=Ni, Pd, Pt), for example, undergo dissociation to form reactive three and two-coordinate species.

The ease of oxidative addition is also determined by the electronic and steric properties of the coordinated ligands. Phosphines, for example, are good σ donor ligands which increase the electron density on the metal and, hence, both its nucleophilicity and tendency to dissociate. Carbon monoxide, on the other hand, is a π acceptor ligand (see Section 2.2) and can decrease electron density on the metal. Thus, $Ni(CO)_2(Ph_3P)_2$ is relatively unreactive and does not readily dissociate due to the balancing effect of the σ-donor and π-acceptor ligands. The steric properties of ligands also play an important role in determining reactivity. In the $M(Ph_3P)_4$ complexes mentioned above, for example, the steric bulk of the four triphenylphosphine ligands is obviously an important factor in determining their ease of dissociation.

From the standpoint of catalysis the stability of the addition product is also important since this determines the ease with which the alkyl metal intermediate undergoes further transformations, such as CO insertion. This explains why rhodium(I) compounds are generally more active catalysts than the corresponding iridium(I) compounds. Thus, iridium(I) complexes more readily undergo oxidative addition, but the addition products are rather stable and further transformations are more difficult than with the corresponding rhodium complexes. In other words, the rate-limiting step in a catalytic process such as the carbonylation of methanol (Equations 44–48) is, in the case of iridum(I) catalysis, not oxidative addition.

Interest in oxidative addition reactions was aroused by the discovery, by Vaska in 1964 [49], that the d^8 iridium complex, $IrCl(CO)(Ph_3P)_2$, reacts readily, under mild conditions, with a wide variety of addends to give stable, octahedral d^6 complexes, e.g. with alkyl halides:

$$
\begin{array}{ccc}
\text{Cl} \diagdown \quad \diagup \text{Ph}_3\text{P} & & \text{Cl} \diagdown \overset{\overset{\displaystyle R}{|}}{} \diagup \text{Ph}_3\text{P} \\
\qquad \text{Ir} \qquad + \text{ RX} \longrightarrow & & \qquad \text{Ir} \\
\text{Ph}_3\text{P} \diagup \quad \diagdown \text{CO} & & \text{Ph}_3\text{P} \diagup \underset{\underset{\displaystyle X}{|}}{} \diagdown \text{CO}
\end{array} \qquad (50)
$$

It was originally thought [50] that these processes involved a concerted, three-centred mechanism (*cis*-insertion). However, it was later found that the oxidative addition of either the *erythro* or *threo* diastereoisomers of PhCHFCHDBr to $IrCl(CO)(PMe_3)_2$ afforded adducts in which racemization had occurred at the reacting carbon centre. Furthermore, the reaction was shown to be inhibited by typical radical scavengers such as galvinoxyl and promoted by small amounts of radical initiators [51, 52]. These observations are compatible with a radical chain mechanism involving the propagation sequence:

$$RX + RIr^{II} \longrightarrow RIr^{III}X + R\cdot \qquad (51)$$

$$R\cdot + Ir^{I} \longrightarrow RIr^{II}, \text{ etc.} \qquad (52)$$

Oxidative additions of reactive alkyl halides, such as methyl iodide, benzyl chloride and allyl halides to the same iridium(I) complex were, by contrast, unaffected by radical scavengers. Furthermore, the relative rates observed in a series of alkyl halides, e.g. R=Me $>$ Et $>$ i-Pr $>$ t-Bu are consistent with a mechanism involving S_N2 attack of iridium(I) on the alkyl halide.

$$Ir^I + RX \longrightarrow RIr^{III} + X^- \xrightarrow{\text{fast}} RIr^{III}X \tag{53}$$

It is now generally recognized that oxidative addition reactions proceed by three different mechanisms — concerted nucleophilic displacement S_N2, radical chain and radical nonchain. The choice of a particular reaction pathway depends on a variety of factors such as the nature of the alkyl group, the halogen, the metal and the surrounding ligands. All three mechanisms have been observed in the oxidative addition of alkyl halides to platinum(0) phosphine complexes [53, 54]:

Concerted nucleophilic displacement (S_N2)

$$Pt^0 + RX \longrightarrow RPt^{II} + X^{-\cdot} \longrightarrow RPt^{II}X \tag{54}$$

Radical chain ($S_{RN}1$)

$$Pt^0 + RX \longrightarrow Pt^IX + R\cdot \tag{55}$$

$$Pt^0 + R\cdot \longrightarrow RPt^I \tag{56}$$

$$RPt^I + RX \longrightarrow RPt^{II}X + R\cdot \text{, etc.} \tag{57}$$

Radical nonchain

$$Pt^0 + RX \longrightarrow [Pt^IX + R\cdot] \xrightarrow{\text{fast}} RPt^{II}X \tag{58}$$

The reactions of carbonylmetallate anions with organic halides (see preceding section) may also in some cases involve radical mechanisms [55]. Similarly, many metal carbonyls catalyse the free radical addition of organic halides, such as CCl_4, to olefins by the following steps [56]:

$$M(CO)_n + CCl_4 \xrightarrow{-CO} MCl(CO)_{n-1} + Cl_3C\cdot \tag{59}$$

$$Cl_3C\cdot + RCH{=}CH_2 \longrightarrow R\overset{\cdot}{C}HCH_2CCl_3 \tag{60}$$

$$R\overset{\cdot}{C}HCH_2CCl_3 \xrightarrow[\text{or MCl}]{CCl_4} RCHClCH_2CCl_3 \tag{61}$$

In all of these reactions of metal complexes with organic halides and related compounds the initial step involves competing single electron transfer (*SET*) from metal to organic halide and two-equivalent nucleophilic displacement (see Scheme 2.11). The former would lead to a radical chain or radical nonchain process depending on the fate of the alkyl radical formed in the subsequent cleavage of the radical anion.

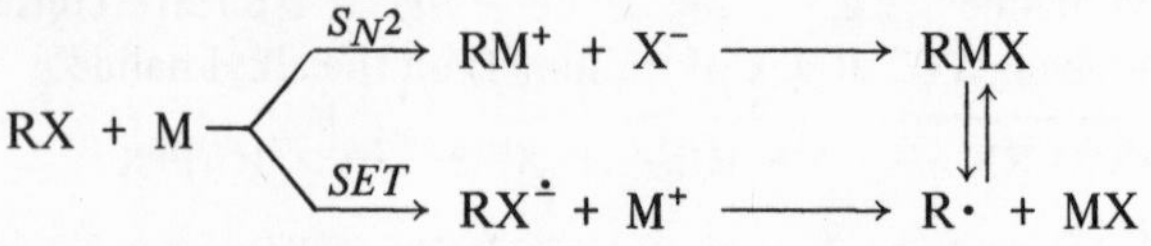

$$RX + M \begin{cases} \xrightarrow{S_N2} RM^+ + X^- \longrightarrow RMX \\ \qquad\qquad\qquad\qquad \updownarrow \\ \xrightarrow{SET} RX^{\cdot-} + M^+ \longrightarrow R\cdot + MX \end{cases}$$

Scheme 2.11

Which pathway is followed appears to depend more on the structure of the alkyl halide, coordinated ligands, solvent, etc., than on the metal employed. The simultaneous occurrence of one-equivalent and two-equivalent processes in many of these systems suggests that the two processes are energetically rather similar. Single electron transfer mechanisms are, moreover, not limited to transition metal chemistry. It is now well established that the formation and reactions of many main group organometallic reagents, such as Grignard reagents, involve single electron transfers and radical intermediates [57], e.g.

$$RX + Mg^0 \longrightarrow [RX^{\cdot-} + Mg^+] \longrightarrow$$
$$R\cdot + MgX \longrightarrow RMg^{II}X \tag{62}$$

The question of homolytic vs. heterolytic carbon–halogen bond cleavage in oxidative additions of alkyl halides to metal complexes is actually a special case of a more general mechanistic problem. The latter is concerned with the question of single electron transfer (*SET*) versus two equivalent nucleophilic and electrophilic substitutions in organic [58–60] and organometallic chemistry [5, 61]. Thus, certain nucleophilic substitutions in aromatic [58] and aliphatic [59] systems have been shown to proceed via a radical chain mechanism, referred to as the $S_{RN}1$ mechanism. The reaction constitutes an overall nucleophilic substitution of Y for X by the following propagation steps.

$$RX^{\cdot-} \longrightarrow R\cdot + X^- \tag{63}$$

$$R\cdot + Y^- \longrightarrow RY^{\cdot-} \tag{64}$$

$$RY^{\cdot-} + RX \longrightarrow RY + RX^{\cdot-} \tag{65}$$

This reaction scheme is mechanistically equivalent to the radical chain mechanism for oxidative addition described in Equations 55–57 (as may readily be seen by replacing Y^- with Pt^0 in Equations 63–65).

Oxidative addition is not restricted to alkyl halides but also occurs with aryl, vinyl and acyl halides as well as with related compounds such as anhydrides, esters, etc. [45–48]. The reductive elimination of an acyl halide is a product-forming step in the catalytic carbonylation of methanol (see Scheme 2.10). The oxidative addition of acetic anhydride to rhodium(I) is a key step in the rhodium-catalysed conversion of acetic anhydride to acetaldehyde and vinyl acetate (see Chapter 7).

$$Rh^I + (CH_3CO)_2O \longrightarrow CH_3CORh^{III}O_2CCH_3 \qquad (66)$$

Summarizing, the oxidative addition of molecular hydrogen and alkyl halides to low-valent, Group VIII metal complexes constitutes a useful method of forming M—H and M—alkyl bonds, respectively. These are key, substrate-activation steps in many catalytic reactions of CO and H_2. Product formation occurs in the following step which usually involves migratory ligand insertion.

2.5 Migratory Insertion and Elimination

In transition metal-catalysed processes substrate activation via coordination and oxidative addition is generally followed by migratory ligand insertion. In this process two ligands coordinated to the same metal centre undergo mutual interaction as shown in the general equation:

$$
\begin{array}{c}
X \\
| \\
M-Y
\end{array}
\rightleftharpoons M-X-Y \qquad (67)
$$

These processes are generally reversible as shown and the reverse process constitutes an elimination of the group X from the coordinated substrate XY. In the context of syn gas chemistry there are two important insertion processes: carbon monoxide insertion and olefin insertion.

CARBON MONOXIDE INSERTION

Carbon monoxide readily undergoes intramolecular insertion into metal—carbon sigma bonds as indicated in the general equation:

$$\qquad (68)$$

The reaction involves migration of the alkyl group to the carbon of the CO ligand and is, therefore, more correctly described as an alkyl migration. The reaction is reversible and the driving force for reaction is generally provided by coordination of a ligand L, which in many cases is CO, at the vacant site created by ligand migration.

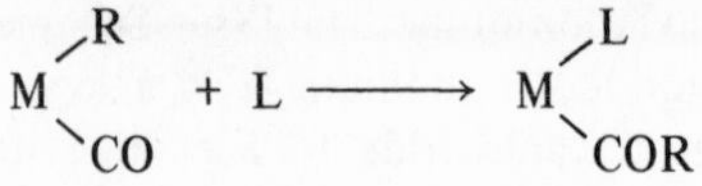

$$M\overset{\displaystyle R}{\underset{\displaystyle CO}{\diagdown}} + L \longrightarrow M\overset{\displaystyle L}{\underset{\displaystyle COR}{\diagdown}} \qquad (69)$$

The reverse process is the key step in metal-catalysed decarbonylations. CO insertion into metal hydride bonds (R=H in Equation 69) is thermodynamically much less favourable than the corresponding reaction with alkylmetal bonds. Consequently, formylmetal complexes are thermally unstable and readily undergo decarbonylation. They are, however, widely believed to be transitory intermediates in the catalytic hydrogenation of carbon monoxide (see Section 2.6).

The insertion process has been difficult to study with the catalytically active cobalt and rhodium complexes involved in many syn gas reactions due to the labile nature of the acylmetal intermediates. Mechanistic studies of CO insertion have, therefore, generally been carried out with alkylpentacarbonylmanganese(I) complexes $RMn(CO)_5$ which are less reactive. In benzene solution, under an atmosphere of carbon monoxide, methylpentacarbonylmanganese(I) is converted to acetylpentacarbonylmanganese(I) [62].

$$CH_3Mn(CO)_5 + CO \rightleftharpoons CH_3COMn(CO)_5 \qquad (70)$$

Increasing the carbon monoxide pressure drives the equilibrium to the right. When the reaction is carried out in the presence of ^{13}CO no labelled carbon is found in the acetyl group of the product, demonstrating that the methyl group migrates to a CO already coordinated to the metal. The reaction proceeds through a 5-coordinate intermediate to which the incoming CO ligand adds as shown below.

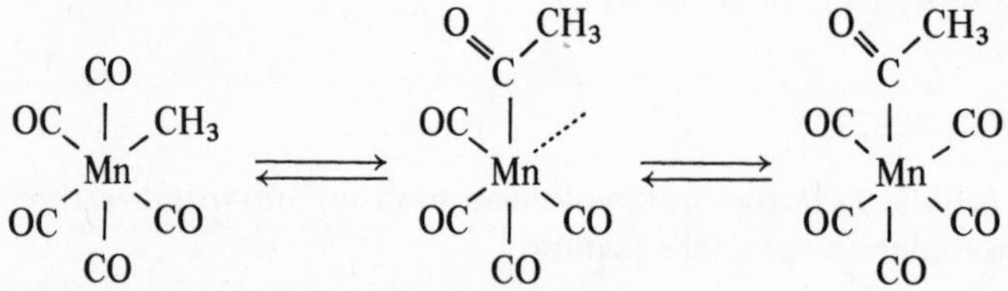

$$(71)$$

CO insertion is a key step in a wide variety of cobalt- and rhodium-catalysed reactions of CO such as olefin hydroformylation and methanol carbonylation and homologation (see Chapters 4 and 7).

OLEFIN INSERTION

Olefin insertion is a ubiquitous reaction in the catalytic chemistry of olefins. Insertion of an olefin into a metal hydride bond, for example, is a key step in metal-catalysed hydrogenations and hydroformylations.

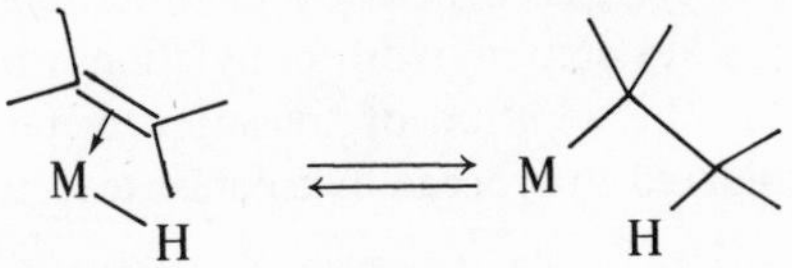

$$\text{[structure] M} \overset{}{\underset{}{\rightleftharpoons}} \text{M [structure]} \tag{72}$$

Here again the reaction is better described as a migration of the hydride ligand to the carbon atom of the coordinated olefin. The reverse reaction, β-hydride elimination, in which a β-hydrogen of an alkyl metal species migrates to the metal centre, is a commonly observed decomposition pathway for alkyl-metal complexes. Indeed, the ease of β-hydrogen elimination is responsible for the instability of many alkyl metal compounds. Alkyl derivatives such as methyl, benzyl and neopentyl, which contain no β-hydrogen, are more stable than, for example, the ethyl derivatives because the lowest energy pathway for decomposition is not available [63]. Evidence has been presented that the presence of a vacant coordination site at the metal is a requirement for facile β-hydrogen elimination. Thus, thermal decomposition of $(n\text{-Bu})_2\,Pt(Ph_3\,P)_2$ is inhibited by the presence of excess triphenylphosphine indicating that dissociation of a phosphine ligand is necessary before β-hydrogen elimination can occur [64].

The metal-catalysed isomerisation of olefins generally involves sequential olefin insertion and β-hydrogen elimination as shown below.

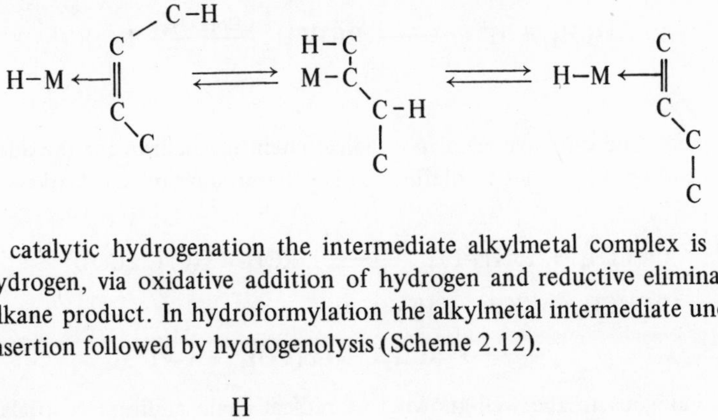

$$\tag{73}$$

In catalytic hydrogenation the intermediate alkylmetal complex is cleaved by hydrogen, via oxidative addition of hydrogen and reductive elimination of the alkane product. In hydroformylation the alkylmetal intermediate undergoes CO insertion followed by hydrogenolysis (Scheme 2.12).

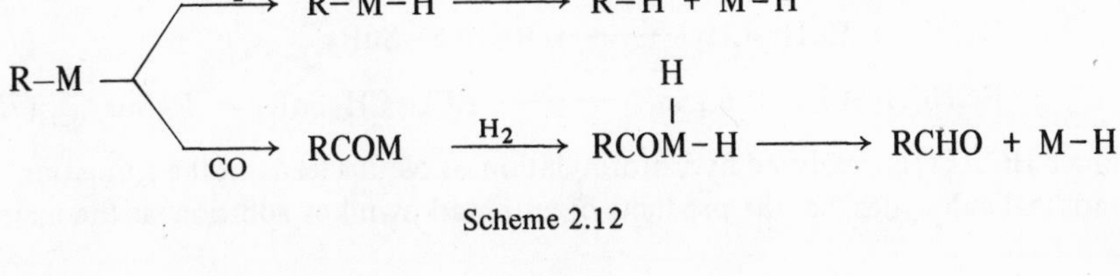

$$R\text{--}M \begin{cases} \xrightarrow{H_2} R\text{--}\overset{\overset{\displaystyle H}{|}}{M}\text{--}H \longrightarrow R\text{--}H + M\text{--}H \\[2em] \xrightarrow[CO]{} RCOM \xrightarrow{H_2} RCO\overset{\overset{\displaystyle H}{|}}{M}\text{--}H \longrightarrow RCHO + M\text{--}H \end{cases}$$

Scheme 2.12

A reaction of particular relevance to syn gas chemistry is the insertion of olefins into metal carbonyl hydrides. It is a key step in hydroformylations catalysed by $HCo(CO)_4$ and related complexes. The addition of transition metal carbonyl hydrides to olefins is generally considered to proceed by a concerted, *cis*-insertion mechanism [38].

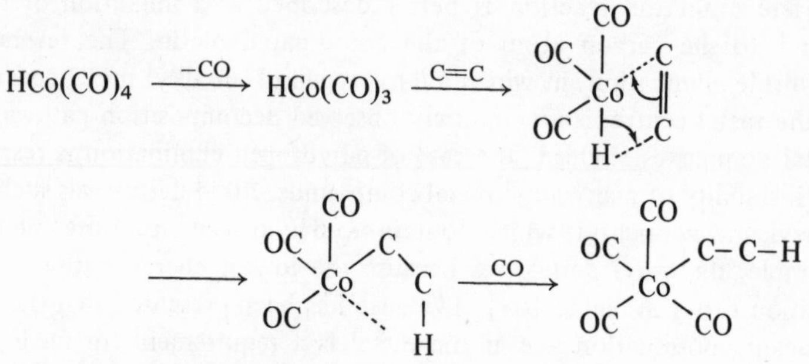

$$HCo(CO)_4 \xrightarrow{-CO} HCo(CO)_3 \xrightarrow{C=C}$$

Scheme 2.13

A priori, one might expect a strong acid such as $HCo(CO)_4$ (see earlier) to add to olefins by a stepwise ionic mechanism. Terminal olefins should then give the product of Markownikov addition:

$$RCH{=}CH_2 + H^+ \longrightarrow R\overset{+}{C}HCH_3 \xrightarrow{Co(CO)_4^-} \underset{\underset{Co(CO)_4}{|}}{RCHCH_3} \qquad (74)$$

Alternatively, one can envisage a free radical chain mechanism for the addition of metal carbonyl hydrides to olefins to give the product of anti-Markownikov addition.

$$Co(CO)_4 + RCH{=}CH_2 \longrightarrow R\overset{.}{C}H{-}CH_2{-}Co(CO)_4 \qquad (75)$$

$$R\overset{.}{C}HCH_2Co(CO)_4 + HCo(CO)_4$$

$$\longrightarrow RCH_2CH_2Co(CO)_4 + Co(CO)_4 \qquad (76)$$

This is analogous to the well-known free radical chain addition of trialkyltin hydrides [43] or HBr to olefins, e.g.

$$R_3Sn\cdot + RCH{=}CH_2 \longrightarrow R\overset{.}{C}HCH_2SnR_3 \qquad (77)$$

$$R\overset{.}{C}HCH_2SnR_3 + R_3SnH \longrightarrow RCH_2CH_2SnR_3 + R_3Sn\cdot \qquad (78)$$

Since $HCo(CO)_4$-catalysed hydroformylation of olefins leads to the formation of terminal aldehydes, i.e. the products of anti-Markownikov addition, as the major products, ionic addition can be ruled out as a major pathway for reaction. Evidence for free radical intermediates in the analogous reaction of $HMn(CO)_5$ with olefins has been presented [65]. By analogy with the addition of HBr to olefins one might expect both ionic and free radical addition to occur, depending on the solvent polarity, the presence of radical initiators, etc. In this context it is worth mentioning that anti-Markownikov addition in cobalt-catalysed hydroformylations is favoured by the addition of phosphine ligands. This is usually attributed to the steric effect of the phosphine ligands but it is worth noting that phosphine ligands also significantly decrease the acidity of the $HCo(CO)_4$ catalyst and, hence, the rate of ionic addition.

We conclude that more detailed mechanistic studies of the reaction of metal carbonyl hydrides, such as $HCo(CO)_4$, with olefins are needed in order to ascertain the importance of free radical pathways.

2.6 Carbon Monoxide Hydrogenation Pathways

The catalytic hydrogenation of carbon monoxide involves the reaction of CO and H_2 at a transition metal centre. This can result in the formation of hydrocarbons or oxygenates, the two simplest examples being the conversion of syn gas to methane and methanol, respectively.

$$CO + 3H_2 \longrightarrow CH_4 + H_2O \tag{79}$$

$$CO + 2H_2 \longrightarrow CH_3OH \tag{80}$$

On the other hand, reaction can result in both reduction and carbon–carbon bond formation, i.e. reductive oligomerization and polymerization. Examples are the Fischer–Tropsch synthesis of hydrocarbons and the conversion of syn gas to ethylene glycol:

$$n\ CO + (2n+1)H_2 \longrightarrow C_nH_{2n+2} + n\ H_2O \tag{81}$$

$$2\ CO + 3\ H_2 \longrightarrow HOCH_2CH_2OH \tag{82}$$

HYDROCARBON FORMATION VIA CARBIDE INTERMEDIATES

A superficial examination of the voluminous literature on CO hydrogenation indicates that the type of reaction observed depends on whether the catalyst is heterogeneous or homogeneous. Soluble transition metal complexes appear to catalyse the formation of oxygenates whilst hydrocarbon formation is observed on metal surfaces, i.e. with heterogeneous catalysts. However, a closer inspection of the literature reveals that hydrocarbon formation is observed with heterogeneous metal catalysts, such as Ni, Fe, Co, Ru, that are able to adsorb carbon monoxide dissociatively to form surface carbide species. Indeed, evidence has been presented by several authors [66–70] that surface carbides are the initially formed intermediates in both methanation (Reaction 79) and the Fischer–Tropsch synthesis (Reaction 81). For example, Araki and Ponec [68] observed that in the hydrogenation of CO over clean nickel surfaces at low pressures the evolution of CO_2 preceded that of CH_4 consistent with the following reaction sequence:

$$2 \ CO \longrightarrow C_{ads} + CO_2 \tag{83}$$

$$C_{ads} + 2 H_2 \longrightarrow CH_4 \tag{84}$$

Similarly, Biloen and coworkers [66, 67] have recently shown that ^{13}C predeposited on a Fischer–Tropsch catalyst becomes incorporated in the hydrocarbon products during reaction with ^{12}CO and H_2 in a manner consistent with the intermediacy of metal carbides. Recent studies of Brady and Pettit [71] have demonstrated that chain propagation in the Fischer–Tropsch reaction involves polymerization of surface methylene (CH_2) species, formed presumably by reduction of surface carbide (see Chapter 3 for a more detailed discussion).

FORMATION OF OXYGENATES VIA FORMYLMETAL INTERMEDIATES

Heterogeneous metal catalysts that do not readily dissociate CO catalyse the formation of oxygenates. The most notable example is the heterogeneous copper catalyst used in methanol synthesis (see Chapter 6). The hydrogenation of CO over other metals, e.g. Pd, Pt and Ir, that do not readily dissociate CO has also been shown to selectively produce methanol [72].

Similarly, soluble transition metal complexes are also unable to dissociate CO and the catalytic hydrogenation of CO with homogeneous catalysts leads to the formation of oxygenates. The primary intermediate in all of these reactions is most probably a formyl metal complex formed by insertion of CO into a metal-hydride bond.

$$\underset{\underset{CO}{|}}{M-H} \ \rightleftharpoons \ M-\overset{\overset{O}{\|}}{C}-H \tag{85}$$

All of the oxygen-containing products observed in the catalytic hydrogenation of CO in the presence of soluble metal complexes can be explained on the basis of initial formation of a formylmetal complex (see below). The feasibility of hydride attack on coordinated CO has been demonstrated in a number of model systems. Thus a wide variety of anionic formylmetal complexes has been prepared by treating metal carbonyls with reducing agents such as trialkoxyboro-hydrides [73, 74], e.g.

$$L_n M(CO) + [HB(OR)_3]^- \longrightarrow [L_n M(CHO)]^- + (RO)_3 B \qquad (86)$$

$$L = Ph_3 P, CO \; ; \; M = Fe, Cr, W$$

A number of stable formylmetal complexes have now been characterised [73–81]. These complexes generally decompose thermally to the metal hydride and carbon monoxide (the reverse of Reaction 85). In one instance the direct formation of a formylmetal complex by reaction of CO with a transition metal hydride has been reported. Thus, rhodium(III)octaethylporphyrin hydride, Rh(OEP)H, reacts with CO in benzene at ambient temperature to give the corresponding formyl complex [81].

$$Rh^{III}(OEP)H + CO \; \rightleftharpoons \; Rh^{III}(OEP)CHO \qquad (87)$$

$$(I)$$

The formyl complex (I) is indefinitely stable in the solid state or in benzene solution in the presence of a CO atmosphere. When (I) is dissolved in benzene, slow decomposition to $[RhOEP]_2$ occurs, via Rh(OEP)H, indicating the reversibility of formylmetal complex formation. The authors suggested, without further discussion, that Reaction (87) may involve free radical intermediates. One can envisage a radical chain mechanism involving a paramagnetic Rh^{II} species as the chain transfer agent in the propagation steps:

$$Rh^{II} + CO \longrightarrow Rh^{III}-\overset{\cdot}{C}{=}O \qquad (88)$$

$$Rh^{III}-\overset{\cdot}{C}{=}O + Rh^{III}-H \longrightarrow Rh^{III}-CHO + Rh^{II} \qquad (89)$$

FORMALDEHYDE AS A KEY INTERMEDIATE

The next step in the catalytic hydrogenation of CO is most probably hydrogenolysis of the formylmetal intermediate to formaldehyde (Reaction 90). Although the formation of formaldehyde is thermodynamically unfavourable it has been argued by Fahey [82] that the concentration of formaldehyde permitted by thermodynamics is more than sufficient for a transient intermediate. Its intermediacy is supported by rate studies, model reactions of formaldehyde with syn gas and the trapping of formaldehyde during the course of the reaction [82].

$$M-CHO + H_2 \longrightarrow H-\overset{\overset{\displaystyle H}{\displaystyle |}}{M}-CHO \longrightarrow M-H + H_2 CO \qquad (90)$$

All of the products observed in homogeneously catalysed hydrogenation of CO can be rationalised on the basis of formaldehyde as a key intermediate. The most thoroughly studied homogeneous systems are those based on catalysis by carbonyl complexes of cobalt [82–84], rhodium [85, 86] and ruthenium [87–90]. All of these systems can lead to the formation of methanol or ethylene glycol and derivatives depending on the reaction conditions. The reaction pathways for the formation of these products are delineated in Scheme 2.14 [82].

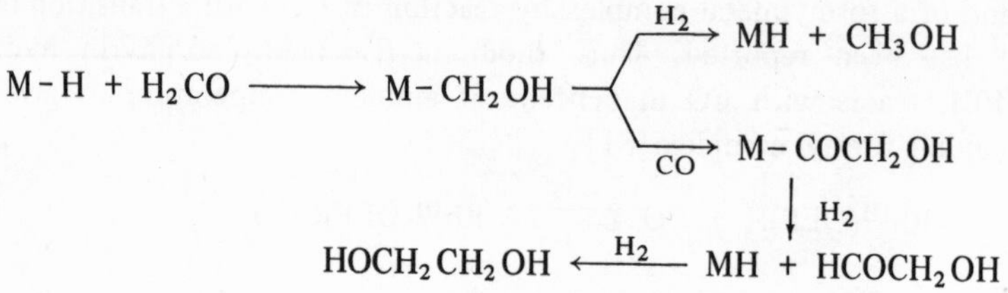

$$M-H + H_2CO \longrightarrow M-CH_2OH \underset{CO}{\overset{H_2}{\big<}} \begin{array}{l} MH + CH_3OH \\ M-COCH_2OH \end{array}$$

$$\qquad\qquad\qquad\qquad\qquad\qquad\qquad\qquad\qquad \downarrow H_2$$

$$HOCH_2CH_2OH \xleftarrow{\ H_2\ } MH + HCOCH_2OH$$

Scheme 2.14

Methanol is formed by simple hydrogenation of formaldehyde. Since the polarization of the M–H bond in metal carbonyl hydrides such as $HCo(CO)_4$ is M^-H^+ it is reasonable to invoke an $M-CH_2OH$ intermediate, formed through attack of M^- at the carbon of the carbonyl group. The formation of ethylene glycol can then be explained by postulating competing CO insertion into the putative $M-CH_2OH$ intermediate, giving glycolaldehyde, and subsequent hydrogenation (see Scheme 2.14). There is good evidence to support such a mechanism. Thus, both cobalt [82, 91] and rhodium [92] carbonyls have been shown to catalyse the hydroformylation of paraformaldehyde. With cobalt carbonyl glycolaldehyde was the major product at 110°C and ethylene glycol at 160°C [91]. The stoichiometric reaction of $HCo(CO)_4$ with monomeric formaldehyde at 0°C and one bar CO gave, after hydrolysis, glycolaldehyde in 60–90% yield [93].

$$CH_2O + HCo(CO)_4 \xrightarrow{\ CO\ } [(CO)_n CoCOCH_2OH] \xrightarrow{\ H^+\ } HOCH_2CHO \quad (91)$$

According to Scheme 2.14 the ratio of methanol to ethylene glycol should be influenced by the CO pressure as is observed in practice. Thus, ethylene glycol becomes a major product only at very high pressures (see Chapter 9). In the model hydroformylation of formaldehyde in the presence of rhodium carbonyl catalysts the ratio of glycol aldehyde to methanol was also shown to be markedly dependent on the solvent [92]. Hydroformylation is favoured only in *N,N*-disubstituted amide solvents which was attributed to an electronic effect of coordinated amide [92]. The problem of the high pressures required for ethylene glycol formation have been partially overcome by carrying out the ruthenium carbonyl-catalysed reaction of syn gas in acetic acid as solvent [87, 88]. Thus, the formation of ethylene glycol diacetate via Reaction 92 is thermodynamically more favourable than ethylene glycol formation via Reaction 82.

$$2\,CO + 3\,H_2 + 2\,HOAc \longrightarrow (CH_2OAc)_2 + 2H_2O \qquad (92)$$

The glycolaldehyde intermediate formed in Scheme 2.14 can also undergo hydroformylation, in a similar fashion to formaldehyde, to give glycerol. The latter is observed as a byproduct in these systems.

$$HOCH_2CHO + HCo(CO)_3 \longrightarrow HOCH_2\underset{\underset{OH}{|}}{C}HCo(CO)_3$$

$$\overset{CO}{\longrightarrow} HOCH_2\underset{\underset{OH}{|}}{C}HCOCo(CO)_3 \overset{H_2}{\longrightarrow} HOCH_2CH(OH)CHO$$

$$\overset{H_2}{\longrightarrow} HOCH_2CH(OH)CH_2OH$$

Scheme 2.15

Other products observed in these systems are formate esters, ethanol [94], higher alcohols and acetaldehyde [95]. A typical product distribution observed by Feder and Rathke [84] in the $HCo(CO)_4$-catalysed hydrogenation of CO at 238 bar and $182°C$ is shown in Table 2.II.

TABLE 2.II
Product distribution in $HCo(CO)_4$-catalysed hydrogenation
of CO at 238 bar and $182°C$

Product	Mole %	
	Dioxane solvent	Aqueous dioxane solvent (16/84% v)
CH_3OH	19.3	2.4
C_2H_5OH	21.8	26
n-C_3H_7OH	5.9	18
n-C_4H_9OH, CH_3CHO	0.9	5.2
$HOCH_2CH_2OH$	26.3	43
Formates	15.6	—
CH_4	4.2	5.4
CO_2	0.05	11

Taken from Feder and Rathke [84].

The formation of methyl formate was postulated [82, 84] to occur via insertion of CO into the methoxycobalt intermediate (V) (see Scheme 2.16). It was suggested [84] that the formation of (V) and the isomeric hydroxymethyl cobalt complex (VI) involves a common intermediate, namely a formaldehyde—cobalt complex (IV). As shown in Scheme 2.16 intramolecular migration of hydride to the coordinated formaldehyde can give (V) or (VI) by addition to carbon or oxygen, respectively. It was noted [84] that the observed products can be adequately explained by means of a formaldehyde—cobalt complex without requiring the formation of free formaldehyde as postulated by Fahey [82]. The formaldehyde complex may, however, be in equilibrium with free formaldehyde, as shown in the scheme.

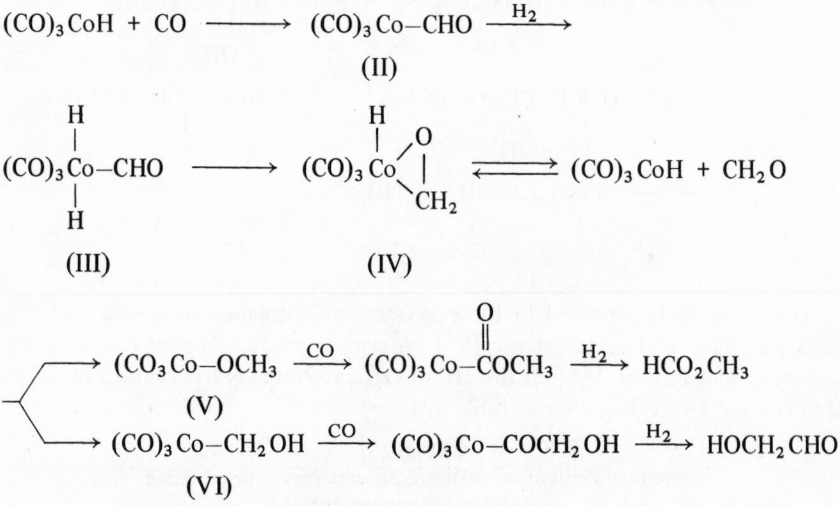

$$(CO)_3CoH + CO \longrightarrow (CO)_3Co-CHO \xrightarrow{H_2}$$

$$(II)$$

$$\begin{matrix} H \\ | \\ (CO)_3Co-CHO \\ | \\ H \end{matrix} \longrightarrow (CO)_3Co\overset{H}{\underset{CH_2}{\overset{|}{\diagup}}}\diagup{}^{O} \rightleftharpoons (CO)_3CoH + CH_2O$$

$$(III) \hspace{4cm} (IV)$$

$$\longrightarrow (CO_3Co-OCH_3 \xrightarrow{CO} (CO)_3Co-\overset{O}{\overset{\|}{C}}OCH_3 \xrightarrow{H_2} HCO_2CH_3$$

$$(V)$$

$$\longrightarrow (CO)_3Co-CH_2OH \xrightarrow{CO} (CO)_3Co-COCH_2OH \xrightarrow{H_2} HOCH_2CHO$$

$$(VI)$$

Scheme 2.16

A stable dihapto formaldehyde complex of osmium, (VII), has recently been fully characterised [96]. Interestingly, (VII) rearranges to the formylhydride complex (VIII) on heating to 75°C. This is the reverse of the pathway postulated for formaldehyde formation in the hydrogenation of CO (see Reaction 90).

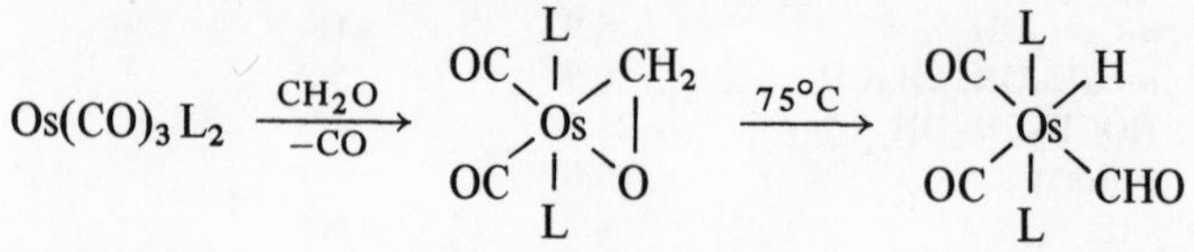

$$Os(CO)_3L_2 \xrightarrow[-CO]{CH_2O} \underset{(VII)}{\overset{L}{\underset{L}{\overset{|}{\underset{OC}{\overset{OC}{\diagdown}}}\overset{|}{\underset{|}{Os}}\overset{CH_2}{\underset{O}{\diagup}}}} \xrightarrow{75°C} \underset{(VIII)}{\overset{L}{\underset{L}{\overset{|}{\underset{OC}{\overset{OC}{\diagdown}}}\overset{|}{\underset{|}{Os}}\overset{H}{\underset{CHO}{\diagup}}}}$$

$$L=Ph_3P$$

A precedent for the reduction of coordinated CO to a coordinated methoxy group is the reported hydrogenation of the zirconium complex (IX) to the methoxy complex (X), which yields methanol on hydrolysis [97].

$$(\eta^5\text{-}C_5Me_5)_2Zr(CO)_2 + 2\,H_2 \xrightarrow[\text{or } 110°C]{h\nu} (\eta^5\text{-}C_5Me_5)_2Zr(H)OCH_3 + CO$$

$$\text{(IX)} \qquad\qquad\qquad\qquad\qquad\qquad\qquad \text{(X)}$$

The formation of formate esters in the $HCo(CO)_4$-catalysed hydrogenation of CO was not observed when water was deliberately added to the system (see Table 2.II). The disappearance of formates was accompanied by the appearance of a virtually equivalent amount of carbon dioxide. This was rationalised [84] on the basis of hydrolysis of the formate esters to formic acid followed by $HCo(CO)_4$-catalysed decarboxylation of the latter.

The formation of ethanol and higher alcohols in this system involves the $HCo(CO)_4$-catalysed homologation of methanol (see Chapter 7). The mechanism of this reaction is well-established (see Chapter 7) and involves the carbonylation of an intermediate methylcobalt complex to give acetaldehyde as the initial product (Scheme 2.17). The latter is rapidly hydrogenated to ethanol under the reaction conditions. The ethanol, in turn, can undergo $HCo(CO)_4$-catalysed homologation to give n-propanol. The small amounts of methane observed arise from hydrogenolysis of the methylcobalt intermediate.

$$CH_3OH + HCo(CO)_4 \xrightarrow{-H_2O} CH_3Co(CO)_4 \xrightarrow{CO} CH_3COCo(CO)_4$$

with branch: $\xrightarrow{H_2} CH_4 + HCo(CO)_4$ from $CH_3Co(CO)_4$ (XI)

$$CH_3COCo(CO)_4 \xrightarrow{H_2} HCo(CO)_4 + CH_3CHO \xrightarrow{H_2} CH_3CH_2OH$$

Scheme 2.17

As discussed earlier (see Section 2.3) the formation of the key intermediate (XI) from methanol in Scheme 2.17 proceeds via nucleophilic displacement of water from protonated methanol by $Co(CO)_4^-$ (see Equation 34). This reaction is facile because of the strongly acidic character of $HCo(CO)_4$. It also explains why methanol homologation is not observed as a reaction pathway in the hydrogenation of CO catalysed by ruthenium complexes as this metal does not form a strongly acidic carbonyl hydride [98].

HOMOGENEOUS VS. HETEROGENEOUS CATALYSIS

As we have seen in the above discussion the hydrogenation of CO in homogeneous systems in the presence of carbonyl complexes of cobalt, rhodium and ruthenium leads to the formation of oxygenates. The wide variety of products observed can be rationalised on the basis of formylmetal complexes and formaldehyde (free or coordinated) as reactive intermediates. Hydrocarbon formation, on the other hand, occurs with heterogeneous metal catalysts that are able to adsorb CO dissociatively to form surface carbides as intermediates. The fundamental difference between homogeneous and heterogeneous catalysis in these systems is readily illustrated in the case of cobalt. Thus, in their study of the $HCo(CO)_4$-catalysed hydrogenation of CO, Feder and Rathke [84] noted that cobalt precipitation sometimes occurred at long reaction times. In experiments where this occurred, significant yields of a Fischer–Tropsch distribution of straight-chain alkanes were observed. Similarly, homogeneous ruthenium carbonyl catalysts afford oxygenates [87–90] whilst hydrocarbon formation is observed in the same system under conditions where ruthenium metal precipitates [99].

The chain growth mechanisms for the formation of oxygenates and hydrocarbons are fundamentally different. The former proceeds by a sequence of classical organometallic processes: CO activation, oxidative addition, migratory insertion and reductive elimination (see Scheme 2.18). The chain growth mechanism currently favoured for hydrocarbon formation is one involving the insertion of a surface-bound methylene into a surface alkylmetal species. The surface methylene is derived from hydrogenation of surface carbide and is probably bonded to two metal centres. Hydrogenation affords a surface methyl complex that can undergo an insertion reaction with another methylene. This reaction would seem to require the simultaneous involvement of 3 metal centres.

Oxygenate formation:

$$M-CHO \longrightarrow MH(CH_2O) \longrightarrow M-CH_2OH \xrightarrow{CO} M-\overset{\overset{\textstyle O}{\|}}{C}CH_2OH, \text{ etc.}$$

Hydrocarbon formation:

$$M-\overset{..}{C}-M \xrightarrow{H_2} M-CH_2-M \xrightarrow{H_2} M-CH_3 \xrightarrow{\text{``}CH_2\text{''}} M-CH_2CH_3, \text{ etc.}$$

Scheme 2.18

Metal cluster complexes [100, 101] constitute a sort of transition phase between soluble, mononuclear transition metal complexes and metal surfaces. Thus, although metal cluster complexes afford homogeneous solutions the close proximity of several metal atoms in these complexes closely parallels the situation pertaining on metal surfaces. It is instructive, therefore, to examine the reactions of CO/H_2 mixtures in the presence of metal carbonyl clusters as models for heterogeneous systems. Muetterties and coworkers have shown that the iridium and osmium clusters, $Ir_4(CO)_{12}$ and $Os_3(CO)_{12}$, catalyse the hydrogenation of CO to alkanes in molten salts, such as $NaAlCl_4$, at $170-180°C$ and $1-2$ bar [102]. Reaction in toluene solution led to the selective formation of methane [103].

Steinmetz and Geoffroy [104] were able to prepare the methyleneosmium carbonyl cluster complex (XIII) by acidification of the formyl complex (XII). Complex (XII) contains a methylene group bonded to two osmium atoms and constitutes a model for the putative surface methylene invoked as an intermediate in hydrocarbon formation over metal surfaces. When (XIII) was heated in an atmosphere of hydrogen methane was formed (Scheme 2.19).

$$Os_3(CO)_{12} + [(RO)_3BH]^- \longrightarrow [Os_3(CO)_{11}CHO]^-$$
$$(XII)$$

$$\xrightarrow{H^+} [Os_3(CO)_{11}(=CHOH)] \xrightarrow{(XII)} [Os_3(CO)_{11}CH_2OH]^-$$

$$\longrightarrow Os_3(CO)_{11}CH_2 + H_2O$$
$$(XIII)$$

Scheme 2.19

Such reductions of coordinated CO to hydrocarbons are not restricted to polynuclear cluster complexes. Thus, an excellent example of the reduction of CO in a mononuclear complex is the $NaBH_4$ reduction of (XIV) which, under carefully controlled conditions, affords successively the formyl, hydroxymethyl and methyl complexes [77].

$$[CpRe(CO)_2NO]^+ + NaBH_4 \xrightarrow[0°C, 15\ min]{THF/H_2O} CpRe(CO)(NO)CHO$$
$$(XIV)$$

$$\begin{array}{c} NaBH_4 \\ THF \end{array} \Bigg\downarrow \begin{array}{c} 0°C, \\ 30\ min \end{array} \qquad\qquad \begin{array}{c} NaBH_4 \\ THF/H_2O \end{array} \Bigg\downarrow \begin{array}{c} 0°C, \\ 15\ min \end{array}$$

$$CpRe(CO)(NO)CH_3 \xleftarrow[25°,\ 5\ hr]{NaBH_4} CpRe(CO)(NO)CH_2OH$$

(Cp = cyclopentadienyl ligand)

Similarly, stoichiometric reduction of Group VI metal carbonyls with AlH_3 [105] or Cp_2NbH_3 [106, 107] affords ethylene as a primary product. A likely precursor of ethylene would seem to be the methylidene complex which could afford ethylene by a 2+2 cycloaddition and subsequent elimination.

$$\begin{array}{c} M{=}CH_2 \\ M{=}CH_2 \end{array} \longrightarrow \begin{array}{c} M{-}CH_2 \\ | \quad | \\ M{-}CH_2 \end{array} \longrightarrow \begin{array}{c} M \\ \| \\ M \end{array} + \begin{array}{c} CH_2 \\ \| \\ CH_2 \end{array} \qquad (93)$$

Gladysz and coworkers [108] have recently isolated and characterised the electrophilic rhenium–methylidene complex (XV):

$$CpRe(Ph_3P)(NO)CH_3 \xrightarrow[CD_2Cl_2,\ -70^\circ C]{Ph_3C^+PF_6^-} \begin{array}{c} CpRe^+(Ph_3P)(NO) \\ \| \\ CH_2 \end{array} \qquad (94)$$

(XV)

These authors pointed out that both heterogeneous and homogeneous hydrogenation catalysts often contain acidic (electrophilic) components such as silica supports, metal oxides, and $AlCl_3$, that could facilitate the formation of a transient methylidene complex. A likely mode of formation of such an intermediate in CO hydrogenation would seem to be the elimination of water from a hydroxymethyl complex as shown below.

$$M{-}CH_2OH + H^+ \longrightarrow M{-}CH_2{-}\overset{+}{O}H_2 \longrightarrow \overset{+}{M}{=}CH_2 + H_2O \qquad (95)$$

or

$$\begin{array}{c} M{-}CH_2OH \\ | \\ H \end{array} + H^+ \longrightarrow \begin{array}{c} M{-}CH_2{-}\overset{+}{O}H_2 \\ | \\ H \end{array} \xrightarrow{-H^+} M{=}CH_2 + H_2O \qquad (96)$$

The proton source in metal-catalysed CO hydrogenations could be an acidic support or cocatalyst or a metal carbonyl hydride which, in many cases, is strongly acidic (e.g. $HCo(CO)_4$). We note, however, that further studies are necessary in order to ascertain the relevance of the stoichiometric reductions discussed above to the mechanism of hydrocarbon formation in the Fischer–Tropsch process. These studies do demonstrate that hydrocarbon formation is, in principle, feasible with mononuclear transition metal complexes and does not always have to involve metal–carbide bond formation.

Similarly, the formation of oxygenates does not always have to involve catalysis by mononuclear transition metal complexes. Thus, many polynuclear metal carbonyls are able to catalyse oxygenate formation from CO/H_2. The most extensively studied example is the catalytic hydrogenation of CO to ethylene glycol in the presence of the cluster anion $Rh_{12}(CO)_{30}{}^{2-}$ [85]. It should be noted, however, that it is often very difficult to define what is actually the active catalyst in these systems because of facile, reversible dissociation to mononuclear species.

2.7 Reactions of CO/H$_2$O

As we noted earlier (see Section 2.3) the reaction of metal carbonyls with water in basic media leads to the formation of metal carbonyl hydrides. Since the latter are reactive intermediates in the catalytic hydrogenation of CO it is possible to carry out many typical reactions of CO/H$_2$ with CO/H$_2$O mixtures. For example, Reppe and Vetter [109] reported an interesting variation of the hydroformylation reaction in which CO/H$_2$O was used in the presence of Fe(CO)$_5$ as catalyst in basic, aqueous medium.

$$RCH{=}CH_2 + 3\,CO + 2\,H_2O \xrightarrow{Fe(CO)_5} RCH_2CH_2CH_2OH + 2\,CO_2 \qquad (97)$$

The disadvantage of this process is that it consumes two extra moles of carbon monoxide compared to the analogous syn gas reaction. It has the advantage, however, that it proceeds under much milder conditions (100°C and 13 bar) than the normal hydroformylation reaction with cobalt catalysts (see Chapter 4).

Reaction (97) and analogous reactions have been studied extensively by the groups of Pettit [110] and Laine [111–113]. *A priori* one might expect that the mechanism of the Reppe modification of the hydroformylation reaction involves the *in situ* generation of hydrogen via the water gas shift reaction (see Section 2.3) followed by a normal hydroformylation pathway. Such a pathway is, however, not tenable since Fe(CO)$_5$ is a very poor catalyst in the hydroformylation reaction with CO/H$_2$. Pettit and coworkers [110] proposed the mechanism shown in Scheme 2.20.

Catalyst generation:

$$Fe(CO)_5 + HO^- \longrightarrow [(CO)_4Fe{-}CO_2H]^- \xrightarrow{-CO_2} [(CO)_4FeH]^-$$

$$\underset{HO^-}{\overset{H_2O}{\rightleftarrows}} H_2Fe(CO)_4 \rightleftarrows H_2Fe(CO)_3 + CO$$

Hydroformylation:

$$RCH{=}CH_2 + H_2Fe(CO)_3 \longrightarrow RCH{=}CH_2 \xrightarrow{CO} RCH_2CH_2Fe(CO)_4H$$
$$\downarrow$$
$$H_2Fe(CO)_3$$

$$\xrightarrow{HO^-} [RCH_2CH_2Fe(CO)_4]^- \longrightarrow [RCH_2CH_2COFe(CO)_3]^-$$

$$\xrightarrow[CO]{H_2O} RCH_2CH_2COFe(CO)_4H \xrightarrow{CO} RCH_2CH_2CHO + Fe(CO)_5$$

Reduction:

$$RCH_2CH_2CHO \xrightarrow{HFe(CO)_4^-} RCH_2CH_2\overset{\overset{\displaystyle O^-}{\displaystyle |}}{CH}-Fe(CO)_4H$$

$$\xrightarrow{H_2O} RCH_2CH_2\overset{\overset{\displaystyle OH}{\displaystyle |}}{CH}-Fe(CO)_4H \xrightarrow{CO}$$

$$RCH_2CH_2CH_2OH + Fe(CO)_5$$

Scheme 2.20

Hydroformylation was observed only when the pH of the solution was reduced to 10.7, i.e. under conditions where the anion $[HFe(CO)_4]^-$ is converted to the carbonyl hydride, $H_2Fe(CO)_4$. The latter is the hydroformylation catalyst. The high activity of $Fe(CO)_5/H_2O$ compared to the poor activity of $Fe(CO)_5/H_2$ is a consequence of the fact that the active catalyst $H_2Fe(CO)_4$ is much more readily formed in the former system.

This system can also effect the hydrogenation of carbon monoxide to methanol. Thus, the reaction of aqueous potassium carbonate with $Fe(CO)_5$ at 100°C and 20 bar CO pressure afforded methanol, together with larger amounts of hydrogen and formate ion [110]. It was suggested that methanol was formed via the formyliron complex (XVI). Interestingly, formation of the latter was postulated to occur by intermolecular nucleophilic attack of $[HFe(CO)_4]^-$ on $Fe(CO)_5$, followed by reductive elimination (Scheme 2.21).

$$Fe(CO)_5 + [HFe(CO)_4]^- \longrightarrow [(CO)_4Fe-\overset{\overset{\displaystyle O}{\displaystyle ||}}{C}-Fe(CO)_4H]^-$$

$$\xrightarrow{-Fe(CO)_4} [(CO)_4Fe-CHO]^- \xrightarrow{H^+} (CO)_4Fe(H)CHO$$

$$(XVI)$$

$$\longrightarrow CH_2O \xrightarrow{[HFe(CO)_4]^-} CH_3OH$$

Scheme 2.21

This system is not effective for the *catalytic* hydrogenation of CO to methanol due to the concommitant formation of formate ion which consumes stoichiometric amounts of base.

2.8 Summary

The wide variety of reactions observed with syn gas in the presence of homogeneous transition metal catalysts can readily be understood on the basis of a series of fundamental organometallic processes: ligand substitution, molecule activation, oxidative addition and reductive elimination, carbon monoxide insertion and β-hydride elimination. In order to be an active catalyst a particular complex must be capable of undergoing all of these steps at a reasonable rate. This is a characteristic property of the Group VIII metals, in particular the low-valent complexes of the first and second row elements: iron, ruthenium, cobalt, rhodium, nickel and palladium.

As a result of the enormous effort devoted in recent years to understanding the mechanism of CO hydrogenation, a clear picture is beginning to emerge of these complex processes. The formation of oxygenates involves the classical organometallic processes mentioned above and formylmetal complexes as the key intermediates.

Hydrocarbon formation over heterogeneous catalysts, on the other hand, appears to involve the surface polymerization of reactive methylidenemetal species ($M{=}CH_2$ or $M{-}CH_2{-}M$). There is much evidence in favour of metal carbides, formed by dissociative adsorption of carbon monoxide, being the precursors of the putative methylidenemetal intermediates. The formation of the latter from formylmetal precursors in some cases cannot, however, be ruled out.

Despite the enormous advances booked in recent years in understanding the mechanisms of syn gas reactions in general and CO hydrogenation in particular there are still many details which remain to be clarified. In particular, the question of homolytic, one-electron processes versus heterolytic two-electron processes in many of these systems would seem to require more study.

References

1. C. A. Tolman, P. Z. Meakin, D. L. Lindner and J. P. Jesson, *J. Am. Chem. Soc.*, **96**, 2762 (1974).
2. R. Ugo, *Chim. Ind.* (Milan), **51**, 1319 (1969).
3. C. A. Tolman, *Chem. Soc. Rev.*, **1**, 337 (1972).
4. J. Halpern, in *Fundamental Research in Homogeneous Catalysis*, Vol. 3, (M. Tsutsui, Ed.), Plenum Press, New York, 1979, p. 25.
5. J. K. Kochi, *Organometallic Mechanisms and Catalysis*, Academic Press, New York, 1978.
6. J. R. Graham and R. J. Angelici, *Inorg. Chem.*, **6**, 2082 (1967).
7. M. S. Wrighton and M. A. Schroeder, *J. Am. Chem. Soc.*, **95**, 5764 (1973); **98**, 551 (1976).
8. M. Absi-Halabi and T. L. Brown, *J. Am. Chem. Soc.*, **99**, 2982 (1977).

9. D. J. Cox and R. Davis, *Inorg. Nucl. Chem. Lett.*, **13**, 301 (1977).
10. B. H. Byers and T. L. Brown, *J. Am. Chem. Soc.*, **97**, 947 (1975); **99**, 2527 (1977).
11. R. Eisenberg and D. E. Hendriksen, *Advan. Catal.*, **28**, 78 (1979).
12. W. Hieber and F. Leutert, *Z. Anorg. Allg. Chem.*, **204**, 145 (1932).
13. P. C. Ford, *Acc. Chem. Res.*, **14**, 31 (1981).
14. P. C. Ford, R. G. Rinker, R. M. Laine, C. Ungerman, V. Landis and S. A. Maya, *Advan. Chem. Ser.*, **173**, 81 (1979).
15. R. M. Laine, R. G. Rinker and P. C. Ford, *J. Am. Chem. Soc.*, **99**, 252 (1977).
16. H. Kang, C. H. Mauldin, T. Cole, W. Slegeir, K. Cann and R. Pettit, *J. Am. Chem. Soc.*, **99**, 8323 (1977); R. Pettit, K. Cann, T. Cole, C. H. Mauldin and W. Slegeir, *Advan. Chem. Ser.*, **173**, 121 (1979).
17. A. D. King, R. B. King and D. B. Yang, *J. Am. Chem. Soc.*, **102**, 1028 (1980); C. C. Frazier, R. Hanes, A. D. King and R. B. King, *Advan. Chem. Ser.*, **173**, 94 (1979).
18. T. Yoshida, Y. Ueda and S. Otsuka, *J. Am. Chem. Soc.*, **100**, 3942 (1978).
19. D. J. Darensbourg, B. J. Baldwin and J. A. Froelich, *J. Am. Chem. Soc.*, **102**, 4688 (1980).
20. B. R. James, G. L. Rempel and F. T. T. Ng, *J. Chem. Soc., A*, 2454 (1969).
21. C. H. Cheng, D. E. Hendriksen and R. Eisenberg, *J. Am. Chem. Soc.*, **99**, 2791 (1977).
22. E. C. Baker, D. E. Hendriksen and R. Eisenberg, *J. Am. Chem. Soc.*, **102**, 1020 (1980).
23. T. Yoshida, T. Okano, Y. Ueda and S. Otsuka, *J. Am. Chem. Soc.*, **103**, 3411 (1981).
24. D. F. Shriver, *J. Organometal. Chem.*, **94**, 259 (1975); *Chem. Brit.*, **8**, 419 (1972).
25. J. S. Kristoff and D. F. Shriver, *Inorg. Chem.*, **13**, 499 (1974).
26. N. E. Kim, N. J. Nelson and D. F. Shriver, *Inorg. Chim. Acta*, **7**, 393 (1973); N. J. Nelson, N. E. Kim and D. F. Shriver, *J. Am. Chem. Soc.*, **91**, 5173 (1969).
27. R. B. Peterson, J. J. Stezowski, C. Wan, J. M. Burlitch and R. E. Hughes, *J. Am. Chem. Soc.*, **93**, 3532 (1971).
28. J. M. Burlitch and R. B. Peterson, *J. Organometal. Chem.*, **24**, C65 (1970).
29. J. C. Kotz and C. D. Turnipseed, *J. Chem. Soc., Chem. Commun.*, 41 (1970).
30. G. Broden, T. N. Rhodin, C. Bruckner, R. Benbow and Z. Hutych, *Surf. Sci.*, **59**, 593 (1976).
31. G. R. Eady, B. F. G. Johnson and J. Lewis, *J. Chem. Soc. Dalton*, 2606 (1975).
32. J. Halpern, *Disc. Faraday Soc.*, **46**, 7 (1968); *Advan. Chem. Ser.*, **70**, 1 (1968).
33. B. R. James, in *Homogeneous Hydrogenation*, Wiley, New York, 1973 and references therein.
34. R. A. Sheldon and J. K. Kochi, *Metal Catalysed Oxidations in Organic Chemistry*, Academic Press, New York, 1981, p. 73.
35. T. A. Weil, S. Metlin and I. Wender, *J. Organometal. Chem.*, **49**, 227 (1973).
36. N. W. Hoffman and T. L. Brown, *Inorg. Chem.*, **17**, 613 (1978).
37. B. H. Byers and T. L. Brown, *J. Am. Chem. Soc.*, **97**, 3260 (1975).
38. M. Orchin and W. Rupilius, *Catal. Rev.*, **6**, 85 (1972).
39. R. F. Heck, *Organotransition Metal Chemistry*, Academic Press, New York, 1974, p. 201.
40. W. Hieber and W. Hybel, *Z. Elektrochem.*, **57**, 235 (1953).
41. I. Wender, *Catal. Rev.*, **14**, 97 (1976).
42. B. L. Booth and B. L. Shaw, *J. Organometal. Chem.*, **36**, 363 (1972).
43. H. G. Kuivila, *Acc. Chem. Res.*, **1**, 299 (1968); *Advan. Organometal. Chem.*, **1**, 47 (1964).
44. J. P. Collman, *Acc. Chem. Res.*, **8**, 342 (1975).

45. J. P. Collman, *Acc. Chem. Res.*, **1**, 136 (1968).
46. J. P. Collman and W. R. Roper, *Advan. Organometal. Chem.*, **7**, 54 (1968).
47. J. Halpern, *Acc. Chem. Res.*, **3**, 386 (1970).
48. J. K. Stille and K. S. Y. Lau, *Acc. Chem. Res.*, **10**, 434 (1977).
49. L. Vaska, *Acc. Chem. Res.*, **1**, 335 (1968).
50. See for example, R. G. Pearson and W. R. Muir, *J. Am. Chem. Soc.*, **92**, 5519 (1970).
51. J. S. Bradley, D. E. Connor, D. Dolphin, J. A. Labinger and J. A. Osborn, *J. Am. Chem. Soc.*, **94**, 4043 (1972).
52. J. A. Labinger, A. V. Kramer and J. A. Osborn, *J. Am. Chem. Soc.*, **95**, 7908 (1973).
53. A. V. Kramer, J. A. Labinger, J. S. Bradley and J. A. Osborn, *J. Am. Chem, Soc.*, **96**, 7145 (1974).
54. A. V. Kramer and J. A. Osborn, *J. Am. Chem. Soc.*, **96**, 7832 (1974).
55. See for example, P. J. Krusic, P. J. Fagan and J. San Filippo, *J. Am. Chem. Soc.*, **99**, 250 (1977); R. J. Kinney, W. D. Jones and R. G. Bergman, *J. Am. Chem. Soc.*, **100**, 635 (1978).
56. A. S. Nesmeyanov, R. Kh. Freidlina, E. Ts. Chukovskaya, R. G. Petrova and A. B. Belyavsky, *Tetrahedron*, **17**, 61 (1962).
57. B. J. Schaart, H. W. H. J. Bodewitz, C. Blomberg and F. Bickelhaupt, *J. Am. Chem. Soc.*, **98**, 3712 (1976) and references cited therein.
58. J. F. Bunnett, *Acc. Chem. Res.*, **11**, 413 (1978).
59. N. Kornblum, *Angew. Chem. Int. Ed. Engl.*, **14**, 734 (1975).
60. R. W. Alder, *J. Chem. Soc. Chem. Commun.*, 1184 (1980).
61. J. K. Kochi, *Pure Appl. Chem.*, **52**, 571 (1980).
62. F. Calderazzo, *Angew. Chem. Int. Ed. Engl.*, **16**, 299 (1977) and references therein.
63. R. R. Schrock and G. W. Parshall, *Chem. Rev.*, **76**, 243 (1976).
64. G. M. Whitesides, J. F. Gaasch and E. R. Stedronsky, *J. Am. Chem. Soc.*, **94**, 5258 (1972).
65. R. L. Sweany and J. Halpern, *J. Am. Chem. Soc.*, **99**, 8335 (1977).
66. P. Biloen, J. N. Helle and W. M. H. Sachtler, *J. Catal.*, **58**, 95 (1979).
67. P. Biloen, *Recl. Trav. Chim. Pays-Bas*, **99**, 33 (1980).
68. M. Araki and V. Ponec, *J. Catal.*, **44**, 439 (1976).
69. P. R. Wentrcek, B. J. Wood and H. Wise, *J. Catal.*, **43**, 363 (1976).
70. J. G. Ekerdt and A. T. Bell, *J. Catal.*, **58**, 170 (1979).
71. R. C. Brady and R. Pettit, *J. Am. Chem. Soc.*, **102**, 6181 (1980); **103**, 1287 (1981); see also J. C. Hayes, G. D. N. Pearson and J. N. Cooper, *J. Am. Chem. Soc.*, **103**, 4648 (1981).
72. J. A. Rabo, A. P. Rissh and M. L. Poutsma, *J. Catal.*, **53**, 295 (1978).
73. C. P. Casey and S. M. Neumann, *J. Am. Chem. Soc.*, **98**, 5395 (1976); **100**, 2544 (1978).
74. C. P. Casey, S. M. Neumann, M. A. Andrews and D. R. McAlister, *Pure Appl. Chem.*, **52**, 625 (1980).
75. J. A. Gladysz and W. Tam, *J. Am. Chem. Soc.*, **100**, 2545 (1978); J. A. Galdysz and J. C. Selover, *Tetrahedron Letters*, 319 (1978).
76. C. P. Casey, M. A. Andrews and J. E. Rinz, *J. Am. Chem. Soc.*, **101**, 741 (1979).
77. J. R. Sweet and W. A. G. Graham, *J. Organometal. Chem.*, **173**, C9 (1979).
78. W. K. Wong, W. Tam, C. E. Strouse and J. Gladysz, *J. Chem. Soc. Chem. Commun.*, 530 (1979).
79. J. P. Collman and S. R. Winter, *J. Am. Chem. Soc.*, **95**, 4089 (1973).
80. R. L. Pruett, R. C. Schoening, J. L. Vidal and R. A. Fiato, *J. Organometal. Chem.*, **182**, C57 (1979).

81. B. B. Wayland and B. A. Woods, *J. Chem, Soc. Chem. Commun.*, 700 (1981).
82. D. R. Fahey, *J. Am. Chem. Soc.*, **103**, 136 (1981).
83. J. W. Rathke and H. M. Feder, *J. Am. Chem. Soc.*, **100**, 3263 (1978).
84. H. M. Feder and J. W. Rathke, *Ann. N. Y. Acad. Sci.*, **333**, 45 (1980); H. M. Feder, J. W. Rathke, M. J. Chen and L. A. Curtis, *ACS Symp. Series*, **152**, 19 (1981).
85. R. L. Pruett, *Ann. N. Y. Acad. Sci.*, **295**, 239 (1977).
86. A. Deluzarche, R. Fonseca, G. Jenner and A. Kiennemann, *Erdoel Kohle Erdgas Petrochem. Brennst. Chem.*, **32**, 313 (1979).
87. B. D. Dombek, *J. Am. Chem. Soc.*, **102**, 6855 (1980).
88. J. F. Knifton, *J. Chem. Soc. Chem. Commun.*, 188 (1981); *J. Am. Chem. Soc.*, **103**, 3959 (1981).
89. R. C. Williamson and T. P. Kobylinski, *US Patent* 4,170,605 (1979).
90. J. S. Bradley, *J. Am. Chem. Soc.*, **101**, 7419 (1979).
91. *US Patent* 3,920,753 (1975) to Ajinomoto.
92. A. Spencer, *J. Organometal. Chem.*, **194**, 113 (1980).
93. J. A. Roth and M. Orchin, *J. Organometal. Chem.*, **172**, C27 (1979).
94. H. Hachenberg, F. Wunder, E. I. Leupold and H. J. Schmidt, *Eur. Pat. Appl.* 21,330 (1981) to Hoechst; *CA* **94**, 174317g (1981).
95. J. E. Bozik, T. P. Kobylinski and W. R. Pretzer, *US Patent* 4,239,705 (1980) to Gulf Research and Development.
96. K. L. Brown, G. R. Clark, C. E. L. Headford, K. Marsden and W. R. Roper, *J. Am. Chem. Soc.*, **101**, 503 (1979).
97. J. M. Manriquez, D. R. McAlister, R. D. Sanner and J. E. Bercaw, *J. Am. Chem. Soc.*, **98**, 6733 (1976).
98. R. B. King, A. D. King and K. Tanaka, *J. Mol. Catal.*, **10**, 75 (1980); see also G. Jenner, A. Kiennemann, E. Bagherzadah and A. Deluzarche, *React. Kinet. Catal. Letters*, **15**, 103 (1980).
99. M. J. Doyle, A.P. Kouwenhoven, C. A. Schaap and B. van Oort, *J. Organometal. Chem.*, **174**, C55 (1979).
100. E. L. Muetterties, *Bull. Soc. Chim. Belg.*, **84**, 959 (1975); **85**, 451 (1976).
101. M. Moskovits, *Acc. Chem. Res.*, **12**, 229 (1979); H. F. Schaefer, *Acc. Chem. Res.*, **10**, 287 (1977).
102. C. Demitras and E. L. Muetterties, *J. Am. Chem. Soc.*, **99**, 2796 1977; H. K. Wang, H. W. Choi and E. L. Muetterties, *Inorg. Chem.*, **20**, 2661 (1981); H. W. Choi and E. L. Muetterties, *Inorg. Chem.*, **20**, 2664 (1981).
103. M. G. Thomas, B. F. Beier and E. L. Muetterties, *J. Am. Chem. Soc.*, **98**, 1296 (1976).
104. G. R. Steinmetz and G. L. Geoffroy, *J. Am. Chem. Soc.*, **103**, 1278 (1981).
105. C. van der Woude, J. A. van Doorn and C. Masters, *J. Am. Chem. Soc.*, **101**, 1633 (1979).
106. K. S. Wong and J. A. Labinger, *J. Am. Chem. Soc.*, **102**, 3652 (1980).
107. J. A. Labinger and K. S. Wong, *ACS Symp. Series*, **152**, 253 (1981).
108. J. A. Gladysz, W. A. Kiel, G. Y. Lin, W. K. Wong and W. Tam, *ACS Symp. Series*, **152**, 147 (1981).
109. W. Reppe and H. Vetter, *Justus Liebig's Ann. Chem.*, **582**, 133 (1953).
110. R. Pettit, C. Mauldin, T. Cole and H. Kang, *Ann. N. Y. Acad. Sci.*, **295**, 151 (1977).
111. R. M. Laine, *Ann. N. Y. Acad. Sci.*, **333**, 124 (1980).
112. R. M. Laine, *J. Am. Chem. Soc.*, **100**, 6451 (1978); *J. Org. Chem.*, **45**, 3370 (1980).
113. W. J. Thomson and R. M. Laine, *ACS Symp. Series*, **152**, 133 (1981).

ADDITIONAL READING

P. C. Ford, Ed., *Catalytic Activation of Carbon Monoxide*, ACS Symposium Series, No. 152, 1981.

E. L. Muetterties and J. Stein, 'Mechanistic Features of Carbon Monoxide Hydrogenation Reactions', *Chem. Rev.*, 79, 479 (1979).

E. L. Kugler and F. W. Steffgen, Eds., *Hydrocarbon Synthesis from Carbon Monoxide and Hydrogen*, Advan. Chem. Ser., No. 178 (1979).

G. P. Chiusoli, 'New Aspects of Organic Syntheses Catalysed by Group VIII Metal Complexes', *Pure Appl. Chem.*, 52, 635 (1980).

P. T. Wolczanski and J. E. Bercaw, 'On the Mechanisms of Carbon Monoxide Reduction with Zirconium Hydrides', *Acc. Chem. Res.*, 13, 121 (1980).

CHAPTER 3

HYDROCARBON SYNTHESIS

In this chapter we shall be concerned with the conversion of syn gas and methanol to hydrocarbons. Our primary concern will be the conversion of syn gas/methanol to base chemicals such as lower olefins and ethylene in particular. However, the conversion to basic chemicals is inextricably related to the production of liquid fuels (syn fuels). It is unavoidable, therefore, that any treatment of hydrocarbon synthesis should include a discussion of liquid fuels production using the classical Fischer–Tropsch process or the more recently developed methanol conversion over the ZSM-5 catalyst (see below).

Several approaches to the production of lower olefins from syn gas/methanol can be delineated. The classical approach consists of conversion of syn gas to a complex mixture of hydrocarbons in a conventional Fischer–Tropsch process, followed by steam cracking of appropriate fractions to ethylene, propylene, BTX, etc. This is, however, a rather circuitous route and much research effort has been devoted to the direct conversion of syn gas or methanol to lower olefins. A major goal of this research is to maximize the yield of ethylene and several strategies have been employed to achieve this goal. These different approaches are outlined in Scheme 3.1.

Before going on to discuss the various routes from syn gas/methanol to lower olefins we shall first consider the simplest hydrocarbon synthesis reaction: methanation.

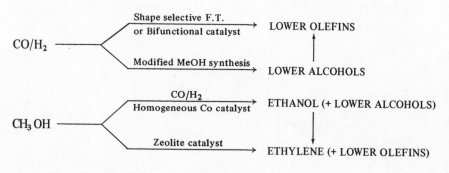

Scheme 3.1

3.1 Methanation

The conversion of syn gas to methane (Reaction 1), over a heterogeneous nickel catalyst at 200–350°C was discovered by Sabatier and Senderens [1] and pre-dates the Fischer–Tropsch process by almost a quarter of a century. The reaction affords a gas containing > 95% methane and constitutes the thermo-dynamically most favourable process for syn gas conversion. Current interest in this reaction stems from the possibility of generating methane, so-called substitute natural gas (SNG), from coal-based synthesis gas.

$$CO + 3\ H_2 \rightleftharpoons CH_4 + H_2O \tag{1}$$

$$\Delta H_{298} = -49.3 \text{ kcal } (-206.4 \text{ kJ})$$

The methane content at equilibrium increases with temperature and commer-cial processes generally operate at temperatures > 280°C and pressures of 20–25 bar. The reverse reaction is the well-known steam reforming of methane over nickel catalysts which is used for generating syn gas (see Chapter 1).

All the Group VIII metals catalyse Reaction 1 to some extent. Although some metals exhibit higher (e.g. Ru) or comparable (e.g. Co) reactivity industrial processes generally employ nickel-based catalysts for reasons of lower price and longer life-time. The catalyst generally consists of nickel supported on alumina or silica often in conjunction with activators such as MgO or Cr_2O_3.

MECHANISM

The methanation reaction is a limiting case of the Fischer–Tropsch conversion of syn gas to a broad range of linear alkanes (see later). As mentioned above Fischer–Tropsch catalysts such as Fe, Co and Ru, are, under the appropriate conditions of temperature and H_2/CO molar ratio, also methanation catalysts. It seems highly likely, therefore, that both of these processes involve a common intermediate: a surface methylene (CH_2) species. Which reaction predominates is determined by the relative rates of hydrogenation and insertion reactions of the methylene intermediate as outlined in Scheme 3.2. These rates are dependent on

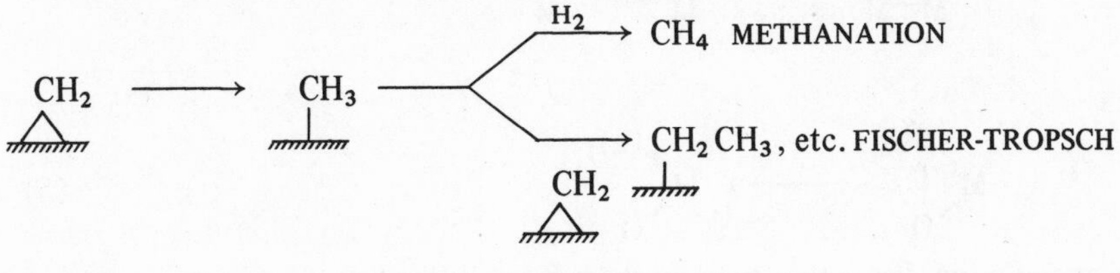

Scheme 3.2

the particular metal employed, the CO and H_2 partial pressures, temperature, space velocity, etc. There is now considerable experimental evidence in support of a methylene insertion mechanism for chain growth in the Fischer–Tropsch process, as is discussed in the next section. One question which remains un-resolved is the origin of the surface methylene species. One school of thought [2, 3] favors initial dissociation of CO into a surface carbide species with con-comitant oxidation of a second molecule of CO to CO_2 as shown:

$$CO_{ads} + CO_s \longrightarrow C_{ads} + CO_2 \tag{2}$$

$$C_{ads} + H_2 \longrightarrow [CH_2]_{ads} \tag{3}$$

One can envisage dissociative chemisorption (Reaction 3) of CO on a metal surface as occurring via cleavage of a bridging CO ligand to a metal carbide and metal oxide, followed by reduction of the latter by a second molecule of CO (as in Scheme 3.3) or, alternatively, by adsorbed hydrogen (i.e. a metal hydride).

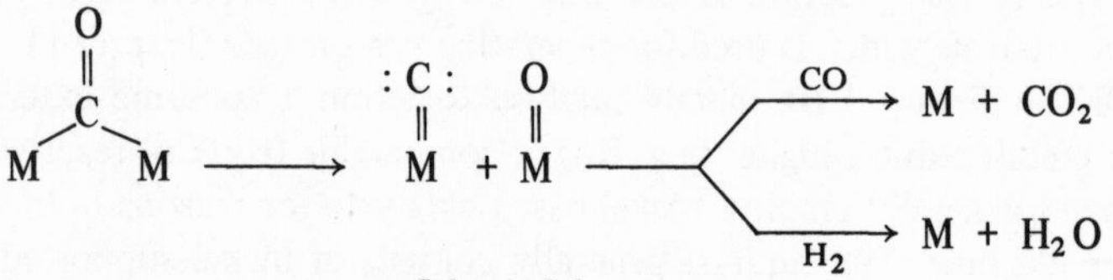

Scheme 3.3

A second school of thought [4] favours stepwise hydrogenation of coordinated (adsorbed) CO via adsorbed formaldehyde (Scheme 3.4)

$$CO_{ads} + H_2 \rightleftharpoons H_2CO_{ads} \xrightarrow{H_2}$$

$$[CH_2OH]_{ads} \longrightarrow CH_{2\,ads} + H_2O$$

Scheme 3.4

Alternatively, one can envisage cleavage of coordinated formaldehyde on the metal surface to give the surface methylene:

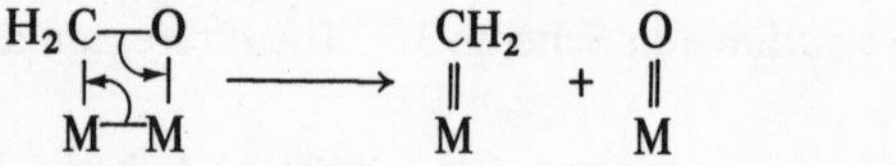

$$\tag{4}$$

Or

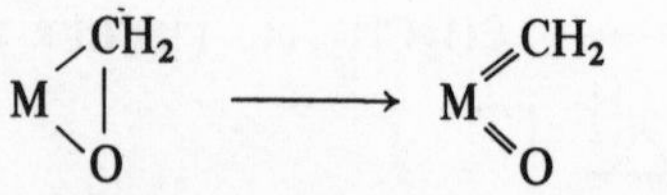

$$\tag{5}$$

An unequivocal distinction between two types of mechanism is not possible at present.

3.2 The Fischer–Tropsch Process

In 1925 Fischer and Tropsch [5] reported the catalytic hydrogenation of carbon monoxide, over an iron-based catalyst at atmospheric pressure, to give a Flory–Schulz distribution of linear alkane and olefins (Reaction 6). The broad spectrum of hydrocarbons extends from methane to hydrocarbon waxes.

$$n \ CO + (2n + 1) \ H_2 \longrightarrow C_nH_{2n + 2} + n \ H_2O \qquad (6)$$

The water formed in reaction (6) can undergo the water gas shift reaction with another molecule of CO:

$$CO + H_2O \longrightarrow CO_2 + H_2 \qquad (7)$$

The overall reaction then becomes:

$$2n \ CO + (n + 1) \ H_2 \longrightarrow C_nH_{2n + 2} + n \ CO_2 \qquad (8)$$

Further research led to the development of cobalt-based catalysts that are superior to iron at atmospheric pressure. The first commercial plants went on stream in Germany in 1936. At the height of production in 1943 the German plants were producing *ca.* 600 000 tons of hydrocarbons, consisting mainly of gasoline (46%) and diesel oil (23%) fractions.

Soon after the Second World War the increasing availability of cheap oil-based hydrocarbons led to the shut-down of the few remaining Fischer–Tropsch plants. The only country to continue with hydrocarbon production by the Fischer–Tropsch process was South Africa. There are basically two reasons for this: an abundant supply of cheap coal and South Africa's wish to be independent of external oil supplies. The SASOL (Suid-Afrikaanse Steenkool Olie en Gaskorporasie) plant came on stream in 1955. The process employs iron-based catalysts at 220°–240°C and 25 bar. Annual production capacity is currently being expanded, with the construction of a SASOL II complex, to around 2 million tons.

Because of the escalating price and the scarcity of oil-based hydrocarbons there has been a revival of interest recently in the Fischer–Tropsch process. A major disadvantage of the process is the broad range of hydrocarbons formed. Recent studies have, therefore, focussed on improving the selectivity to higher grade feedstocks for the chemical industry. This has led to the development of modified catalysts that favour the production of lower olefins (see below).

CATALYSTS

Of all the metals that exhibit significant activity (Fe, Co, Ni and Ru) only iron is currently of importance as it is superior to cobalt with respect to conversion rate, selectivity and flexibility [6]. Nickel catalysts yield mainly methane (see previous section) and ruthenium catalysts yield only high molecular weight alkanes (polymethylene). Other Group VIII metals (Ir, Pd, Pt, Os, Rh) exhibit low activities.

Ruthenium distinguishes itself from the other metals in being an active catalyst at temperatures as low as 100°C. At pressures up to *ca*. 30 bar methane formation predominates. As the pressure increases high molecular weight products become important. The usual pressures employed are in the range 1000–2000 bar. Although this reaction could, in principle, be used as an alternate source of high density polyethylene there are several obstacles to commercialization, such as low selectivity to hydrocarbons of molecular weight > 20 000, low space-time yield and the very high pressures required.

Commercial Fischer–Tropsch catalysts consist of iron supported on silica, alumina or kieselguhr in conjunction with alkali metal salt promotors. Vollhardt and co-workers [7] have recently reported the use of a polystyrene-supported cyclopentadienyl cobalt dicarbonyl catalyst (I) as a Fischer–Tropsch catalyst. Reaction of CO/H_2 (1 : 3) over this catalyst at 140°C and *ca*. 7 bar afforded a typical Flory–Schulz distribution of linear alkanes. Under these conditions (I)

$$(9)$$

$$(I)$$

retained its "homogeneous", mononuclear character during the course of reaction. These results are particularly interesting as they appear to demonstrate the feasibility of catalysis of the Fischer–Tropsch process by a well-defined, mononuclear, organometallic complex. Much recent work has also been devoted to the development of Fischer–Tropsch catalysts comprising metal carbonyls (e.g. Fe, Ru) supported on zeolites. These shape-selective catalysts are discussed further in the following section.

MECHANISM

The mechanism of the Fischer–Tropsch process has been discussed extensively in the literature [8–18]. As we have already mentioned the nature of CO hydrogenation over transition metals varies markedly with the metal employed. Typical Fischer–Tropsch catalysts (Fe, Co, Ru) produce a Flory–Schulz distribution of linear alkanes and olefins. With Ni and Pd methane is the principal product whilst with Cu no reaction is observed.

Since its discovery basically three different mechanisms have been postulated to explain the Fischer–Tropsch reaction. The first, suggested by Fischer and Tropsch in 1926 [19] invoked the formation C–C bonds via the oligomerization of methylene (CH_2) fragments on the metal surface. Recent work of Brady and Pettit [20, 21] has provided convincing evidence in favour of such a mechanism. Thus, a mixture of linear alkanes and olefins was produced when diazomethane (CH_2N_2) was decomposed, at $210°C$ in the presence of hydrogen, over Co, Fe and Ru catalysts [20]. The hydrocarbon distribution observed with a Co catalyst was strikingly similar to that observed with CO/H_2 mixtures over the same catalyst. The results were consistent with a chain growth mechanism involving the insertion of a surface CH_2 species into an alkyl-metal bond to give the next higher homologous alkyl ligand. The polymerization is initiated by metal hydride bonds as shown in Scheme 3.5.

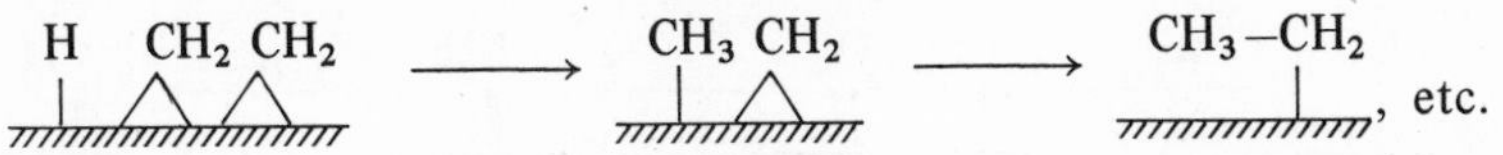

Scheme 3.5

Chain termination can occur through β-hydride elimination of the alkylmetal intermediate to produce a terminal olefin, or by hydrogenolysis of the alkyl-metal to an alkane. In the absence of H_2 the decomposition of CH_2N_2 over Ni, Pd, Fe, Co, Ru and Cu surfaces affords mainly ethylene. It was concluded [20] that the principal reaction of CH_2 fragments alone on a metal surface is dimerization and desorption of the resulting ethylene (Scheme 3.6). Interestingly, the reaction of CH_2N_2 over a Cu surface afforded only ethylene even in the presence of H_2, consistent with the fact that Cu is not a Fischer–Tropsch catalyst. It was noted [20] that Cu does not readily dissociatively chemisorb H_2 to form the metal-hydride bonds, needed for the initiation of CH_2 polymerization.

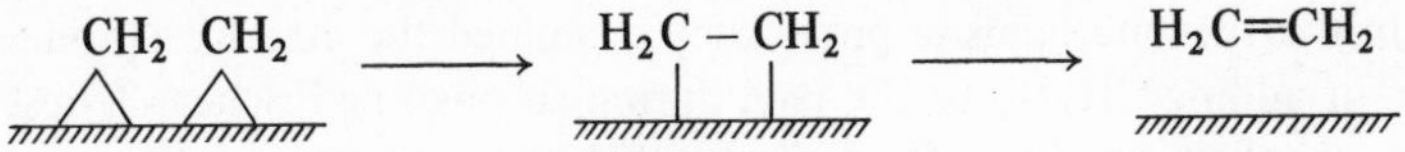

Scheme 3.6

A second mechanism, advanced by Anderson and Emmett and coworkers [14, 15], suggested that C–C bond formation occurs by condensation of surface hydroxymethylidene species (Scheme 3.7).

The most recent mechanism, suggested by Pichler and Schulz [13] and subsequently with minor variations by others [9, 16], invokes C–C bond formation by CO insertion into an alkylmetal bond analogous to the well-known CO insertion reactions encountered in homogeneous systems (see previous chapter).

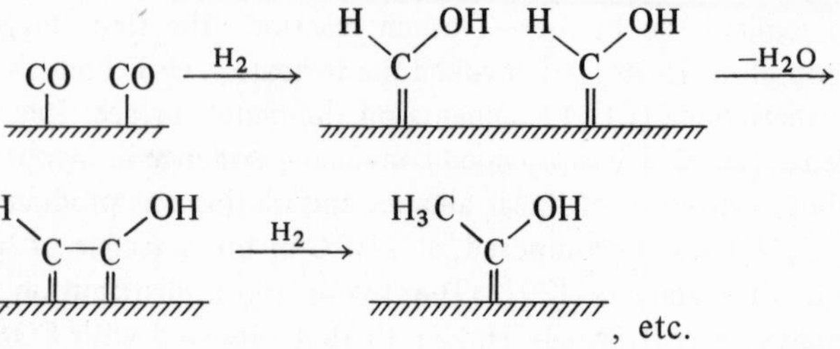

Scheme 3.7

This mechanism is outlined in Scheme 3.8.

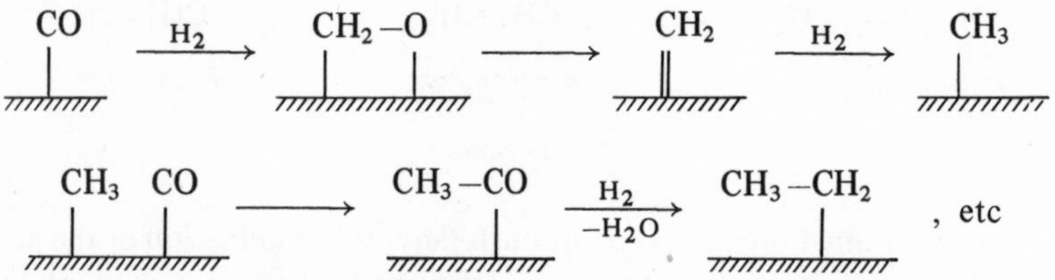

Scheme 3.8

These three schemes differ fundamentally in the role assigned to the putative CH_2 intermediate. The Fischer–Tropsch scheme involves the CH_2 species in both chain initiation and propagation. In the Pichler–Schulz scheme, on the other hand, CH_2 species are involved only as chain initiators whilst in the Anderson–Emmett scheme they play no role at all. Brady and Pettit have recently described [21] experiments designed to test the applicability of these three alternative mechanisms.

In the first mechanistic probe they examined the effect on product distribution of adding CH_2N_2 to the feed during an ongoing Fischer–Tropsch reaction over a cobalt catalyst. They found that the addition of CH_2 groups (from CH_2N_2 decomposition) caused a change in product distribution in favour of higher molecular weight molecules, i.e. the rate of propagation was increased relative to the rate of termination. This result is predicted only by the Fischer–Tropsch scheme.

In the second diagnostic test the distribution of ^{13}C atoms in the propene formed when a mixture of ^{13}CO, H_2 and $^{12}CH_2N_2$ are passed over the cobalt catalyst, was investigated and found to be consistent only with the Fischer–Tropsch scheme. The latter predicts that with pure ^{13}CO the propene should contain every possible combination of ^{13}C and ^{12}C atoms. The Pichler–Schulz mechanism, in contrast, predicts a mixture of $^{13}C-^{13}C-^{13}C$ and $^{12}C-^{13}C-^{13}C$ molecules whilst the Anderson–Emmett scheme predicts a mixture of $^{13}C-^{13}C-^{13}C$ and $^{12}C-^{12}C-^{12}C$ molecules. It was concluded [21], that of the three possible schemes only the original Fischer–Tropsch mechanism accounts for the experimental data.

There are two aspects of the mechanism of the Fischer–Tropsch process that remain to be resolved; the nature and mode of formation of the putative surface methylene intermediate. Brady and Pettit [20, 21] favour a bridging methylene unit between two metal atoms noting that this arrangement is generally observed in model organometallic complexes [22]. However, the observation (see earlier) of a Fischer–Tropsch reaction with the mononuclear cobalt complex (I) would seem to indicate the possibility of a mononuclear, $M=CH_2$ intermediate. As mentioned in the preceding section several pathways can be envisaged for the formation of coordinated CH_2 from coordinated CO and H_2. The currently available experimental data do not allow for unequivocal distinction between these various alternatives.

3.3 Lower Olefins via Modified Fischer–Tropsch Processes

As mentioned earlier a serious drawback of the conventional Fischer–Tropsch process is the lack of selectivity. A major goal of current research on syn gas conversion processes is the selective production of chemical feedstocks. Much effort has been devoted, therefore, towards optimizing the yields of naphtha (C_5-C_{11} alkanes and cycloalkanes) or, preferably, the direct production of lower (C_2-C_4) olefins [23, 24]. The formation of lower olefins is thermodynamically unfavourable compared to that of higher molecular weight, saturated hydrocarbons. It has been noted [23], however, that thermodynamic equilibrium is attained slowly during the reaction, thus allowing for kinetic control of the product distribution.

The product distribution in the Fischer–Tropsch process is influenced by several parameters, such as the particular metal, its particle size, temperature, pressure and CO/H_2 ratio, space velocity, etc. Recent investigations have shown [23] that α-olefins are the primary products and that they are converted to alkanes via subsequent hydrogenation. Due to their high hydrogenation activity cobalt catalysts generally tend to yield mainly saturated hydrocarbons. Iron catalysts, on the other hand, yield substantial amounts of olefins, although a maximum of only 27% lies in the interesting C_2-C_4 range.

In this section we shall be concerned with the significant improvements in selectivity which have been achieved by catalyst modification. There are essentially two problems which have to be solved in order to maximize the production of lower olefins: control of chain growth and inhibition of hydrogenation of olefins to alkanes. Basically three approaches have been used in catalyst modification:

- Highly dispersed catalysts
- Bimetallic catalysts
- Shape selective catalysts

HIGHLY DISPERSED METAL CATALYSTS

The dependence of selectivity on catalyst particle size is readily illustrated by the work of Blanchard and co-workers [25]. These authors showed that a highly dispersed cobalt catalyst, prepared by reduction of cobalt(II) acetylacetonate with triethylaluminium, was active for the conversion of syn gas to lower olefins in an alkylterphenyl solvent at 200°C and atmospheric pressure. No hydrocarbons higher than C_6 were observed and olefins comprised ca. 75% of the C_2–C_6 hydrocarbons. The activity of the catalyst remained stable and the olefin selectivity constantly high over several runs.

Highly dispersed iron catalysts were prepared by Chauvin and co-workers [26] by impregnating inorganic oxide supports, e.g. Al_2O_3, La_2O_3, MgO, SiO_2, with $Fe(CO)_5$, $Fe_3(CO)_{12}$ or $HFe_3(CO)_{11}^-$ followed by thermal decomposition at 180–270°C in an atmosphere of CO/H_2. These catalysts exhibited selectivities towards lower olefins higher than 50%, the product distribution showing a sharp maximum for C_2–C_3 hydrocarbons. The decomposition of the supported anionic cluster $HFe_3(CO)_{11}^-$, for example, produced iron particles with a size smaller than 20 Å. This catalyst exhibited an unusually high selectivity (45%) towards propylene during the first hours of reaction. The selectivity subsequently declined slowly with time and higher molecular weight hydrocarbons were formed.

Subsequent studies [27] showed that ethylene was converted, in 70% yield, to propylene over this catalyst, indicating that ethylene is a primary product of CO/H_2 conversion but is rapidly converted to propylene. It was suggested [27] that the conversion of ethylene to propylene involves a metallocyclobutane intermediate formed as shown in Scheme 3.9. This mechanism is in agreement with model reactions observed with olefins and tantalum alkylidene complexes [28].

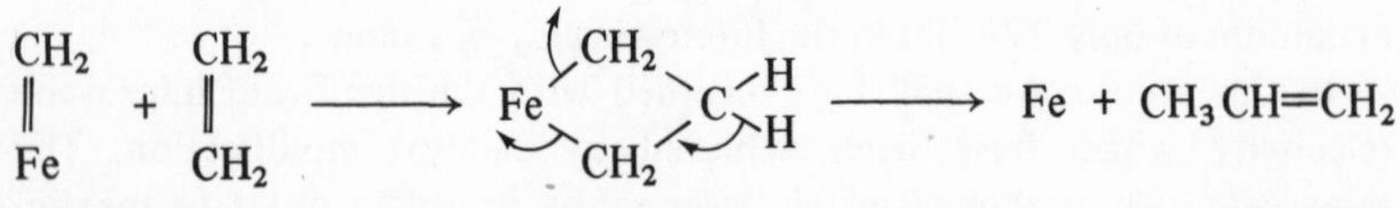

Scheme 3.9

Similarly, higher olefins are formed via homologous metallocyclobutane intermediates, e.g. 1-butene is formed from propylene (Scheme 3.10). This constitutes an interesting explanation of the chain growth mechanism in the Fischer–Tropsch process which, as discussed in the preceding section, has been shown to involve the oligomerization of surface methylene species.

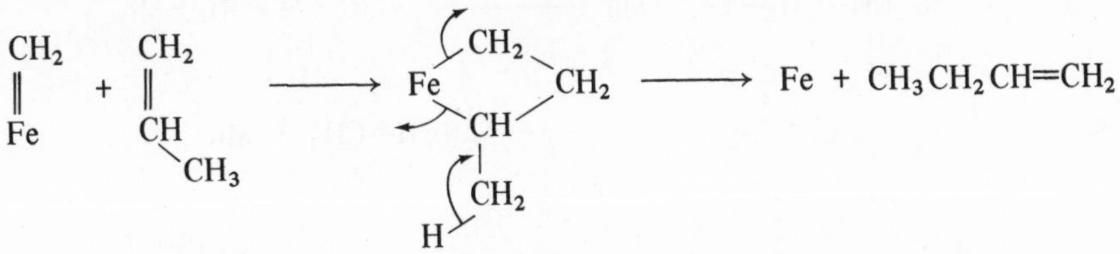

$$CH_2 \quad CH_2 \atop \parallel \quad + \quad \parallel \atop Fe \quad CH \atop \quad \quad \searrow CH_3 \quad \longrightarrow \quad Fe \overset{CH_2}{\underset{CH}{\diagdown}} CH_2 \quad \overset{CH_2}{\underset{H}{\diagup}} \quad \longrightarrow \quad Fe + CH_3CH_2CH{=}CH_2$$

Scheme 3.10

Catalysts prepared from other Group VIII metal carbonyl clusters, e.g. $Rh_4(CO)_{12}$, $Ir_4(CO)_{12}$, $Ru_3(CO)_{12}$ and $Os_3(CO)_{12}$, supported on SiO_2 or Al_2O_3, also favoured the production of low molecular weight hydrocarbons, but in these cases they were mainly saturated [26].

BIMETALLIC CATALYSIS

Increasing the selectivity to lower hydrocarbons should, in principle, be feasible if a second metal is added which suppresses chain growth and ensures desorption of short-chain fragments. Frohning and co-workers [23, 24] found after testing *ca.* 300 different catalyst combinations, that the addition of the oxides of Ti, V, Mo, W and Mn to an iron-based catalyst promoted the formation of olefins and at the same time suppressed the formation of hydrocarbons above C_6. This increase in selectivity was accompanied by a decrease in activity compared to iron alone. This had to be compensated for by increasing the reaction temperature to $> 300°C$ in order to achieve reasonable conversion rates. However, under these conditions the catalyst had low stability due to rapid deactivation via carbon deposition.

Similarly, Kolbel and Tillmetz [29] found that CO hydrogenation over a manganese/iron catalyst (Mn : Fe = 9 : 1) at 330°C resulted in the formation of up to 90% olefinic products consisting largely of ethylene, propylene and butenes.

At the present stage of development the bimetallic, doped iron catalysts appear to have interesting prospects as selective catalysts for the conversion of syn gas to lower olefins. However, the lower activity and short life-time of these catalysts still remains a serious obstacle to their commercialization. Much effort is currently being devoted to further optimization of these catalysts. The precise function of the second metal in these bimetallic systems still remains to be elucidated. It is tempting to speculate that termination of chain growth involves the transfer of an alkyl ligand from iron to the second metal followed by β-hydride elimination to give an olefin, e.g:

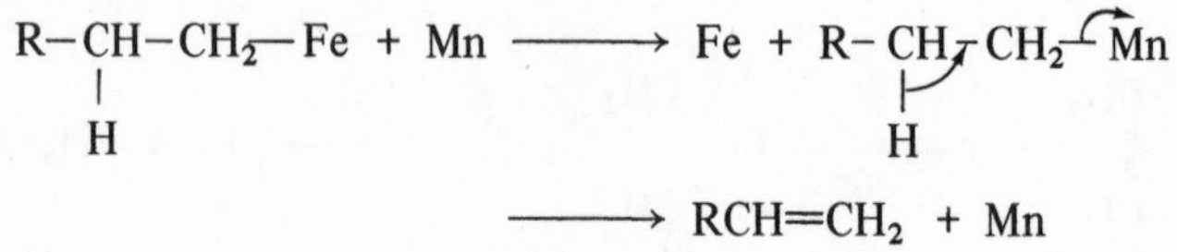

$$R-CH-CH_2-Fe + Mn \longrightarrow Fe + R-CH_2-CH_2-Mn$$
$$\underset{H}{|} \qquad\qquad\qquad\qquad \underset{H}{|}$$

$$\longrightarrow RCH=CH_2 + Mn$$

Scheme 3.11

SHAPE SELECTIVE CATALYSTS

Shape selective catalysis of CO hydrogenation involves the use of transition metal-exchanged zeolites as catalysts. Zeolites, or molecular sieves, are crystalline aluminosilicates consisting of a tetrahedral network of SiO_4 and AlO_4^- units [30, 31]. The negative charges on the AlO_4 tetrahedra are balanced, for example, by hydrated Na^+ ions which are readily exchangeable by other metal ions or protons. The proton-exchanged zeolites (H-form) are strong acids that are widely used in the oil industry as cracking catalysts. Both naturally occurring and synthetic zeolites are known. Their broad potential as catalysts stems from the fact that they possess molecular cavities (zeolite cages) and channels of varying size depending on the zeolite structure. The zeolite cages and channels can, hence, accommodate only molecules of a certain size and geometry and selectivities in hydrocarbon conversion processes can be dramatically influenced by varying the geometry of the cavities and channel dimensions [32]. This has led to the concept of shape selective catalysis [33–35].

Recently a new class of synthetic medium-pore zeolites has been developed that is active for the selective conversion of methanol to aromatic gasoline (see following section). When these so-called ZSM-5 zeolites are exchanged with Group VIII metal ions they afford shape selective CO hydrogenation catalysts [36]. The size of the channels in these zeolites is such that they cannot accommodate hydrocarbons larger than C_{10}. This property is used to advantage in controlling the chain-growth in the Fischer–Tropsch process. Thus, both iron- [37] and ruthenium- [38, 39] exchanged ZSM-5 zeolites have been used for the selective conversion of syn gas to aromatic gasoline. In this process the initially-formed lower olefins are converted to aromatics by an acid-catalysed dehydrocyclization.

The use of related Co and Ru catalysts derived from large-pore zeolites, such as faujasite, has also been reported [40–42]. A further modification is the use of catalysts prepared by deposition of metal carbonyls, such as $Fe_3(CO)_{12}$,

$Co_2(CO)_8$ and $Ru_3(CO)_{12}$ on faujasite for the selective conversion of syn gas to C_1-C_9 hydrocarbons [43]. It is questionable, however, whether the improved selectivities observed with these catalysts derived from large pore zeolites are due to shape selectivity or to the fact that they contain highly dispersed metals (see above).

At the time of writing no report has, to our knowledge, appeared of metal-exchanged zeolite catalysts for the selective conversion of syn gas to lower (C_2-C_4) olefins. However, the recent development of second generation ZSM-type zeolites for the selective conversion of methanol to lower olefins (see following section) would seem to suggest that such a transformation is feasible.

3.4 Methanol Conversion to Hydrocarbons

An alternative to direct conversion of syn gas to hydrocarbons is a two stage conversion via methanol as intermediate. Selective conversion of syn gas to methanol can be achieved using well-established technology (see Chapter 6). The methanol is then converted by a separate step into gasoline or lower olefins.

METHANOL TO GASOLINE

Methanol can be used by itself as an automotive fuel or as a gasoline blend [44]. However, methanol's affinity for water, its corrosiveness, toxicity, lower energy content per unit volume, etc., present serious obstacles to its use as an automotive fuel or gasoline blend. Considering the large investments required to modify engines, storage and distribution facilities, etc., it may be more economical to convert methanol into gasoline.

The non-selective conversion of methanol to a broad range of hydrocarbons can be achieved with a variety of strong acid catalysts. The breakthrough in this area was signalled by the discovery, by Mobil workers [45–50], of a new class of shape-selective zeolites, that catalyse the selective conversion of methanol at 370°C and ca. 15 bar, to a mixture of aliphatic and aromatic hydrocarbons boiling predominantly in the gasoline range (C_5-C_{10}). These synthetic zeolites, called ZSM-5 by the discoverers [45–47], possess a unique channel structure that is intermediate between the familiar wide-pore zeolites, such as faujasite, and the narrow-pore zeolites such as zeolite-A and erionite. Consequently, ZSM-5 excludes molecules with critical dimensions larger than C_{10} aromatics (tetramethylbenzenes), which corresponds to the end point of conventional gasoline. In other words, even if larger molecules could be formed inside the zeolite cage they could not diffuse through the pores and hence would undergo subsequent cracking to smaller molecules.

The distribution of products obtained from the Mobil and SASOL Fischer–

Tropsch processes are compared in Table 3.I. The gasoline produced by the Mobil Process has an unleaded research octane number of 90–95 and is superior in both quality and yield to that produced by the SASOL process. Moreover, in contrast to the SASOL process, the Mobil process requires no elaborate down-stream processing to remove unwanted oxygenates, etc.

TABLE 3.I

Product distributions from the SASOL and Mobil Process[a]

Product		Process	
		SASOL–1	Mobil
Light gas	C_1-C_2	20.1	1.3
LPG	C_3-C_4	23.0	17.8
Gasoline	C_5-C_{12}	39.0	80.9
Diesel oil	$C_{13}-C_{18}$	5.0	0
Heavy oil	C_{19+}	6.0	0
O compounds		7.0	0
Aromatics (% of gasoline)		5.0	38.6

[a]Data taken, with permission, from G. A. Mills, *Chem. Tech.*, 418 (1977).

The Mobil process occurs in three distinct steps as outlined in Scheme 3.12: (a) the dehydration of methanol to dimethyl ether, (b) the dehydration of dimethyl ether to lower olefins, and (c) the transformation of lower olefins to a mixture of aromatics and alkanes [47, 50].

$$2\ CH_3OH \underset{}{\overset{-H_2O}{\rightleftharpoons}} CH_3OCH_3 \xrightarrow{-H_2O} C_2-C_5\ \text{olefins}$$
$$\longrightarrow \text{aromatics} + \text{alkanes}$$

Scheme 3.12

In the process described by Chang and co-workers [46] the reaction is carried out in two stages using dual reactors in fixed-bed operation. The first reactor contains a catalyst that promotes only the dehydration of methanol to dimethyl ether. The second reactor contains the ZSM-5 catalyst that converts both the dimethyl ether and unreacted methanol to hydrocarbons. The benefit of this staging is that the substantial heat of reaction from these highly exothermic reactions can be released under more controlled conditions. The process can utilize crude methanol that often contains up to 25–30% by weight of water.

That the Mobil process involves the sequence of reactions outlined in Scheme 3.12 is readily seen by following the product distribution during reaction as shown in Figure 3.1. The initial dehydration of methanol to dimethyl ether is

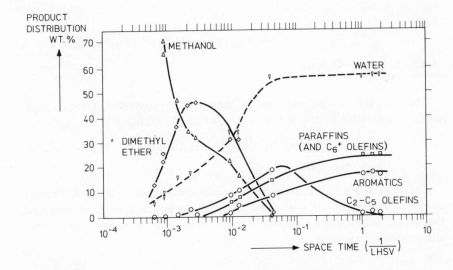

Figure 3.1 Reaction Pathway for Methanol Conversion to Hydrocarbons over ZSM-5 at 371°C (reproduced with permission from Chang and Silvestri [47]).

rapid and reversible with close approach to equilibrium. The subsequent conversion of dimethyl ether to hydrocarbons is rate limiting and is subject to autocatalysis [49], i.e. the rate accelerates rapidly as the concentration of hydrocarbons increases. This observation was rationalised [49] on the basis of the initial rate of formation of the primary product, ethylene, being much slower than its subsequent alkylation by oxygenates.

Little is known about the mechanism of the conversion of dimethyl ether to the primary product, ethylene. Chang and Silvestri [47] favour a pathway involving α-elimination of methanol to methylene (CH_2), followed by insertion of the latter into either methanol or dimethyl ether (Scheme 3.13).

$$CH_3OCH_3 \longrightarrow CH_3OH + :CH_2 \qquad (10)$$

$$:CH_2 + CH_3OCH_3 \longrightarrow CH_3CH_2OCH_3 \qquad (11)$$

$$CH_3CH_2OCH_3 \longrightarrow CH_2{=}CH_2 + CH_3OH \qquad (12)$$

Scheme 3.13

Other authors [51, 52] favour a mechanism involving a zeolite-mediated transfer of an incipient methyl cation from a protonated dimethyl ether molecule to a second molecule of dimethyl ether.

$$H_3C - \overset{+}{\underset{CH_3}{O}} \overset{H}{\diagdown}$$

$$Zeol^- \Big) \Big(\underset{H}{\overset{CH_2-O-CH_3}{\diagup}} \longrightarrow CH_3OH + CH_3CH_2OCH_3 + \text{H-Zeol} \qquad (13)$$

The initially formed ethylene is subsequently converted, through acid-catalysed alkylation reactions, to a mixture of C_3-C_5 olefins, e.g.

$$CH_3OR + CH_2{=}CH_2 \xrightarrow{\text{H-Zeol}} CH_3CH{=}CH_2 + ROH \qquad (14)$$

$$R = \text{H or } CH_3$$

The final step is then acid-catalysed dehydrocyclization of the C_3-C_5 olefins to give a mixture of C_6-C_{10} aromatics. Since no molecular hydrogen is produced the overall stoichiometry requires the co-production of alkanes (and cyclo-alkanes) with the aromatics.

METHANOL TO LOWER OLEFINS OVER ZEOLITES

In the conversion of methanol to gasoline over ZSM-5 type zeolites the primary hydrocarbon products are lower olefins (see above). The selectivity to lower olefins can be significantly improved by adjusting various reaction parameters such as:

space velocity: decreasing the space velocity and, hence, the methanol conversion results in a significant increase in lower olefin selectivities (see Table 3.II).
pressure: lowering the pressure to sub-atmospheric improves lower olefin yields [50].
steam addition: the addition of steam ($> 50\%$ wt on methanol) has a favourable effect [53].
catalyst modification: modification with trimethylphosphite produces a catalyst more selective for olefin production [52].

The most dramatic increase in lower olefin yields has, however, been obtained by using a second generation ZSM zeolite, ZSM-34 [54, 55]. ZSM-34 is a synthetic narrow-pore zeolite of the erionite—offretite family. When *ca.* 25% aqueous methanol is converted over ZSM-34 at 370°C and 1 bar, ethylene is formed in *ca.* 60% selectivity at 72% methanol conversion. The total yield of C_2-C_4 olefins was 89% (see Table 3.II).

A serious drawback of the ZSM-34 catalyst is its very short life-time owing to rapid deactivation through coke deposition. Thus, the catalyst requires regeneration after only a few hours of operation. In addition the synthesis of

TABLE 3.II
Lower olefins from methanol cracking[a]

Catalyst	H–ZSM–5 [b]			H–ZSM–34 [c]		Naphtha cracking[d]
Feed	MeOH	MeOH	MeOH	MeOH	MeOH/H_2O (1 : 3.4)	Naphtha
LHSV	1080	108	1	2	1.2 (MeOH)	
MeOH conv. (%)	9.1	47.5	100	88.2	72.2	
MeOMe sel. (%)	98	69	0	30	5	
Hydrocarbon sel. (%)	2	31	100	70	95	
Hydrocarbon distribution (wt %)						
methane	1.5	1.1	1.1	2.1	1.6	16.0
ethane	–	0.1	0.6	0.4	1.1	3.4
ethylene	18.1	12.4	0.5	42.5	59.7	31.3
propane	2.0	2.5	16.2	1.8	5.2	0.5
propylene	48.2	26.7	1.0	26.1	23.6	12.1
butanes	13.8	7.8	24.3	3.8	1.3	0.2
butenes	11.9	15.8	1.3	6.7	5.8	2.8
C_{5+} alkanes	4.4	27.0	14.0	–	1.4	9.0
aromatics	–	6.6	41.1	16.5	0.3	13.0
C_2H_4 sel. (%)	18.1	12.4	0.5	42.5	59.7	31.3
$C_3^=/C_2^=$ ratio	2.7	2.2	2.0	0.6	0.4	0.4
Total C_2–C_4 olefins (%)	78.2	54.9	2.8	75.3	89.1	47.2

[a] Reactions carried out at 370°C and 1 bar.
[b] Data taken from Chang and Silvestri [47].
[c] Data taken from Givens *et al.* [54].
[d] Data taken from L. F. Hatch and S. Matar, *Hydrocarbon Process.*, 57(1) 135 (1978); and 57(3), 129 (1978).

ZSM-34 is a rather tedious and expensive procedure using choline chloride as crystallization modifier. It is interesting therefore, that Inui and co-workers [56] have recently reported that the use of tetramethylammonium hydroxide as crystallization modifier considerably reduces the crystallization time and affords a more stable catalyst.

Another recent development is the use of manganese-doped zeolites [57]. For example, reaction of 70% aqueous methanol at 400°C and 1 bar over a mixture of manganese-exchanged erionite and chabazite resulted in the formation of a mixture of C_1–C_4 hydrocarbons. At 90% conversion the yields (vol. %) of ethylene and propylene were 37 and 26, respectively. Recycling of the dimethyl ether led to the selectivities shown in Table 3.III.

TABLE 3.III
Methanol to lower olefins over Mn-doped zeolites[a]

Products	Selectivity (%)	
Ethylene	36.3	
Propylene	38.8	80.3%
Butenes	5.2	C_2-C_4 olefins
C_1-C_4 alkanes	19.5	

[a] Taken from Wunder and Leupold [57].

The recent developments described above, and other reports [58] of selective conversion of methanol to lower olefins, would seem to suggest that commercial exploitation of this new technology can be expected in the not too distant future. Commercial viability will probably depend ultimately on finding a compromise between high yield, low coking and stability properties.

BIFUNCTIONAL CATALYSTS

An avenue for possible future study in this area is the development of bifunctional catalyst systems. Thus, combination of the two functions — methanol synthesis and methanol conversion — in a two-component catalyst would, in principle, allow the direct conversion of syn gas to gasoline or lower olefins.

This approach appears to be the underlying theme in the report, by Shell workers [59], of the direct conversion of syn gas to a mixture of aromatic hydrocarbons over a catalyst cocktail comprising a mixture of $ZnO-Cr_2O_3$ and a crystalline gallium silicate. Reaction at $375°C$ afforded a 78% selectivity for C_{5+} (predominantly aromatic) hydrocarbons, at 67% syn gas conversion.

3.5 Ethylene via Methanol Homologation

Another approach to the methanol to ethylene problem is via methanol homologation (Reaction 15) followed by dehydration of the resulting ethanol (Reaction 16).

Reaction 15 is carried out in the liquid phase, at elevated temperatures and pressures, in the presence of homogeneous metal carbonyl catalysts, such as $Co_2(CO)_8$, usually promoted by iodide ion [60, 61]. It is probably one of the most intensively studied reactions in the petrochemical industry at this moment. Recent developments indicate that ethanol selectivities approaching 90% are feasible (see Chapter 7 for a more detailed discussion of methanol homologation).

$$CH_3OH + CO + 2H_2 \longrightarrow CH_3CH_2OH + H_2O \qquad (15)$$

$$CH_3CH_2OH \longrightarrow CH_2{=}CH_2 + H_2O \qquad (16)$$

One could hardly imagine a simpler reaction than the catalytic dehydration of ethanol to ethylene (Reaction 16) the history of which dates back to the 18th century. Nevertheless, recent developments in catalyst and reactor design have resulted in improved economics for this process [62–64]. Industrial processes generally employ either an activated alumina or a supported phosphoric acid catalyst. A fluidized-bed process has been described [63] which affords an ethylene yield of 99%.

Current emphasis on ethanol dehydration also stems from the widespread interest in the production of ethylene from ethanol derived from the fermentation of carbohydrates. In many developing countries – Brazil and India in particular – ethylene production from renewable agricultural resources represents an economically viable alternative to fossil fuel-based ethylene.

3.6 Syn Gas to Lower Olefins via Lower Alcohols

The conversion of syn gas over modified methanol synthesis catalysts can lead to the production of a mixture of lower (C_1–C_4) alcohols (Reaction 17). For example, a catalyst consisting of a mixture of Cu and Co oxides and alkali metal salts, along with the oxides of Cr, Fe, V or Mn, has been described [65, 66]. Conversion of syn gas over this catalyst at 250°C and 60 bar afforded a mixture of linear C_1–C_4 alcohols in > 95% selectivity with ethanol as the major product (38% yield). Subsequent dehydration of the mixture of lower alcohols provides a two-step route from syn gas to lower olefins.

$$CO/H_2 \longrightarrow CH_3OH, \ CH_3CH_2OH, \ CH_3CH_2CH_2OH, \ \text{etc.} \qquad (17)$$

The catalytic synthesis of ethanol from syn gas over supported, pyrolysed rhodium carbonyl clusters has been reported by Ichikawa [67]. The product distribution is markedly dependent on the nature of the support and on the precursor cluster used in catalyst preparation. The formation of ethanol was enhanced on weakly basic oxides, e.g. La_2O_3, TiO_2 and ZrO_2. The best results were obtained with $Rh_4(CO)_{12}/La_2O_3$ (224°C and *ca.* 0.8 bar) which gave 61% ethanol together with 20% methanol as the major products. The mechanisms of these reactions have, as yet, received little attention. It would seem likely that they involve initial formation of methanol followed by methanol homologation (see Chapter 9 for a more detailed discussion of syn gas to lower alcohol conversions).

3.7 Butadiene from Ethanol

It is noteworthy that in the various syn gas/methanol to lower olefin conversions described in the preceding sections little or no butadiene is formed. This contrasts with lower olefin production via naphtha cracking which affords substantial amounts of butadiene (see Chapter 1). Since butadiene is an important industrial chemical, a changeover to syn gas/methanol as feedstock would necessitate finding an alternative source of butadiene. In this context it is worth mentioning that during World War II enormous amounts of butadiene were manufactured from fermentation ethanol in a two-step process:

$$CH_3CH_2OH \longrightarrow CH_3CHO + H_2 \tag{18}$$

$$CH_3CHO + CH_3CH_2OH \longrightarrow CH_2{=}CH{-}CH{=}CH_2 + 2\,H_2O \tag{19}$$

overall reaction: $2\,CH_3CH_2OH$

$$\longrightarrow CH_2{=}CH{-}CH{=}CH_2 + 2\,H_2O + H_2 \tag{20}$$

In the first step ethanol is partially dehydrogenated to acetaldehyde over a copper chromite catalyst. The resulting mixture of ethanol and acetaldehyde is dehydrated at 300–350°C and 1 bar over a Zr or Ta oxide-silica gel catalyst. Overall yields of butadiene are about 70%. Reaction 19 probably involves an initial aldol condensation of two molecules of acetaldehyde.

3.8 Summary

The recent revival of interest in the conversion of syn gas to liquid hydrocarbons stems from the recognized need for alternative routes to basic hydrocarbon building blocks and liquid fuels that are independent of oil as a raw material. In particular, the selective conversion to lower olefins is a desirable goal in the context of chemicals manufacture.

Basically two strategies have been employed to achieve this goal: direct conversion of syn gas to lower olefins and conversion via methanol as an intermediate. The former involves the use of modified Fischer–Tropsch catalysts and the latter involves "methanol cracking" over shape-selective zeolite catalysts. Although these new technologies have yet to be commercialized they show considerable promise and further improvements are constantly being disclosed.

Alternatively, the methanol can be converted to ethanol, by reaction with syn gas (homologation), followed by dehydration. This technology is discussed in more detail in Chapter 7. Yet another approach involves the use of modified methanol synthesis catalysts for the direct conversion of syn gas to lower alcohols, and subsequent dehydration to lower olefins.

References

1. P. Sabatier and J. B. Senderens, *Hebd. Seances Acad. Sci.*, **134**, 514 (1902).
2. M. Araki and V. Ponec, *J. Catal.*, **44**, 439 (1976).
3. P. R. Wentrcek, B. J. Wood and H. Wise, *J. Catal.*, **43**, 363 (1976).
4. M. A. Vannice, *J. Catal.*, **37**, 462 (1975).
5. F. Fischer and H. Tropsch, *German Patent* 484,337 (1925).
6. Y. T. Shah and A. J. Perrotta, *Ind. Eng. Chem. Prod. Res. Dev.*, **15**, 123 (1976).
7. L. S. Benner, P. Perkins and K. P. C. Vollhardt, *ACS Symp. Series*, **152**, 165 (1981).
8. M. A. Vannice, *Catal. Rev.*, **14**, 153 (1976).
9. C. Masters, *Advan. Organometal. Chem.*, **17**, 161 (1979).
10. I. Tkatchenko, in *Fundamental Research in Homogeneous Catalysis*, Vol. 3 (M. Tsutsui, Ed.) Plenum Press, New York, 1979, p. 119.
11. P. Biloen, *Recl. Trav. Chim. Pays-Bas*, **99**, 33 (1980).
12. V. Ponec, *Catal. Rev.*, **18**, 151 (1978).
13. H. Pichler and H. Schulz, *Chem. Ing. Tech.*, **42**, 1162 (1970); H. Schulz, *Erdoel Kohle Erdgas Petrochem. Brennst. Chem.*, **30**, 123 (1977).
14. H. H. Storch, N. Goulombic and R. B. Anderson, *The Fischer–Tropsch and Related Syntheses*, Wiley, New York, 1951.
15. J. Kummer and P. H. Emmett, *J. Am. Chem. Soc.*, **75**, 5177 (1953).
16. G. Henrici-Olivé and S. Olivé, *Angew. Chem. Int. Ed. Engl.*, **15**, 136 (1976).
17. A. T. Bell, *Catal. Rev.*, **23**, 203 (1981).
18. H. Kolbel and M. Ralek, *Catal. Rev.*, **21**, 225 (1980).
19. F. Fischer and H. Tropsch, *Brennst. Chem.*, **7**, 97 (1926); *Chem. Ber.*, **59**, 830 (1926).
20. R. C. Brady and R. Pettit, *J. Am. Chem. Soc.*, **102**, 6181 (1980).
21. R. C. Brady and R. Pettit, *J. Am. Chem. Soc.*, **103**, 1287 (1981).
22. C. E. Sumner, P. E. Riley, R. E. Davis and R. Pettit, *J. Am. Chem. Soc.*, **102**, 1752 (1980) and references cited therein.
23. B. Bussemeier, C. D. Frohning and B. Cornils, *Hydrocarbon Process.*, **55**(11), 105 (1976) and references cited therein.
24. C. D. Frohning in *New Syntheses with Carbon Monoxide*, (J. Falbe, Ed.) Springer-Verlag, Berlin, 1980, p. 356.
25. M. Blanchard, D. Vanhove, F. Petit and A. Mortreux, *J. Chem. Soc. Chem. Commun.*, 908 (1980).
26. D. Commereuc, Y. Chauvin, F. Hugues, J. M. Basset and D. Olivier, *J. Chem. Soc. Chem. Commun.*, 154 (1980); see also T. Okuhara, T. Kimura, M. Misono and Y. Yoneda, *J. Chem. Soc. Chem. Commun.*, 1114 (1981).
27. F. Hugues, B. Besson and J. M. Basset, *J. Chem. Soc. Chem. Commun.*, 719 (1980); see also F. Hugues, B. Besson, P. Bussière, J. A. Dalmon, J. M. Basset and D. Olivier, *Nouv. J. Chim.*, **5**, 207 (1981).
28. S. J. McLain, C. W. Wood and R. R. Schrock, *J. Am. Chem. Soc.*, **99**, 3519 (1977).
29. H. Kolbel and K. D. Tillmetz, *US Patent* 4,177,203 (1979) to Schering.
30. P. B. Venuto, *Chem. Tech.*, **1**, 215 (1971).
31. P. B. Weisz, *Chem. Tech.*, **3**, 498 (1973).
32. P. B. Weisz, V. J. Frilette, R. W. Maatman and E. B. Mower, *J. Catal.*, **1**, 307 (1962).
33. N. Y. Chen and P. B. Weisz, *Chem. Eng. Progr. Symp. Ser.*, **63**, 86 (1967).
34. S. M. Csicsery, *Zeolite Chemistry and Catalysis*, J. A. Rabo, ed., American Chemical Society, Washington, 1976, p. 680.
35. E. Derouane, *Stud. Surf. Sci. Catal.*, **5**, 5 (1980).

36. C. D. Chang, W. H. Lang and A. J. Silvestri, *J. Catal.*, **56**, 268 (1979).
37. P. D. Caesar, J. A. Brennan, W. E. Garwood and J. Ciric, *J. Catal.*, **56**, 274 (1979).
38. T. J. Huang and W. O. Haag, *ACS Symp. Series*, **152**, 307 (1981).
39. W. O. Haag and T. J. Huang, *US Patent* 4,157,338 (1979).
40. H. H. Nijs, P. A. Jacobs and J. B. Uijtterhoeven, *J. Chem. Soc. Chem. Commun.*, 180, 1095 (1979).
41. P. A. Jacobs, J. Verdonck, R. Nijs and J. B. Uijtterhoeven, *Advan. Chem. Ser.*, 178, 15 (1979).
42. D. Fraenkel and B. C. Gates, *J. Am. Chem. Soc.*, **102**, 2478 (1980).
43. D. Ballivet-Tkatchenko, N. D. Chau, H. Mozzanega, M. C. Roux and I. Tkatchenko, *ACS Symp. Series*, **152**, 187 (1981).
44. T. B. Reed and R. M. Lerner, *Science*, **182**, 1299 (1973).
45. S. L. Meisel, J. P. McCullough, C. H. Lechtaler and P. B. Weisz, *Chem. Tech.*, 86 (1976).
46. C. D. Chang, J. W. Kuo, W. H. Lang, S. M. Jacob, J. J. Wise and A. J. Silvestri, *Ind. Eng. Chem. Process. Des. Dev.*, 17, 255 (1978).
47. C. D. Chang and A. J. Silvestri, *J. Catal.*, **47**, 249 (1977).
48. N. Y. Chen and W. E. Garwood, *J. Catal.*, **52**, 453 (1978).
49. N. Y. Chen and W. J. Reagan, *J. Catal.*, **59**, 123 (1979).
50. C. D. Chang, W. H. Lang and R. L. Smith, *J. Catal.*, **56**, 169 (1979).
51. E. Derouane, J. Nagy, P. Dejaifve, J. van Hoof, B. Spekman, J. Vedne and C. Naccache, *J. Catal.*, **53**, 40 (1978).
52. W. W. Kaeding and S. A. Butter, *J. Catal.*, **61**, 155 (1980).
53. L. Marosi, J. Stabenow and M. Schwarzmann, *German Patent* 2,827,385 (1980) to BASF.
54. E. N. Givens, C. J. Plank and E. J. Rosinski, *US Patent* 4,079,095 (1978) to Mobil Oil.
55. E. N. Givens, C. J. Plank and E. J. Rosinski, *US Patent* 4,079,096 (1978) to Mobil Oil.
56. T. Inui, T. Ishihara and Y. Takegami, *J. Chem. Soc. Chem. Commun.*, 936 (1981).
57. F. A. Wunder and E. I. Leupold, *Angew. Chem. Int. Ed. Eng.*, **19**, 126 (1980).
58. A. Kasai, T. Okuhara, M. Misono and Y. Yoneda, *Chem. Letters*, 449 (1981): H. Itoh, T. Hattori and Y. Murakumi, *J. Chem. Soc. Chem. Commun.*, 1091 (1981).
59. M. A. M. Boersma, M. E. M. Post and L. Schaper, *German Patent* 3,027,358 (1981) to Shell.
60. D. W. Slocum, *Catalysis in Organic Syntheses*, (W. H. Jones, Ed.), Academic Press, New York, 1980, p. 245.
61. H. Bahrmann and B. Cornils, *New Syntheses with Carbon Monoxide*, (J. Falbe, Ed.), Springer-Verlag, Berlin, 1980, p. 226.
62. N. K. Kochar and R. L. Marcell, *Chem. Eng.*, Jan. 28, 1980, p. 80.
63. U. Tsao and J. W. Reilly, *Hydrocarbon Process.*, **57**(2), 133 (1978).
64. O. Winter and M. T. Eng. *Hydrocarbon Process.*, **55**(11), 125 (1976).
65. A. Sugier and E. Freund, *German Patent* 2,748,097 (1978) to Institut Français du Pètrole.
66. G. M. Intille, *Div. Petrol. Chem. Preprints, ACS Meeting, Honolulu*, April 7–6, 1979, p. 318.
67. M. Ichikawa, *J. Chem. Soc. Chem. Commun.*, 566 (1978).

ADDITIONAL READING

E. L. Kugler and F. W. Steffgen, Eds. *Hydrocarbon Synthesis from Carbon Monoxide and Hydrogen*, Advan. Chem. Series, No. 178 (1979).

P. C. Ford, Ed., *Catalytic Activation of Carbon Monoxide*, ACS Symp. Series, No. 152 (1981).

L. L. Anderson and D. A. Tilman, *Synthetic Fuels from Coal*, Wiley, New York, 1979.

E. L. Muetterties and J. Stein, 'Mechanistic Features of Catalytic Carbon Monoxide Hydrogenation Reactions', *Chem. Revs.*, 79, 479 (1979).

B. M. Harney and G. A. Mills, 'Coal to Gasoline via Syn Gas', *Hydrocarbon Process.*, 59(2), 67 (1980).

J. Haggin, 'Fischer–Tropsch: New Life for Old Technology', *C. and E. News*, Oct. 26, 1981, p. 22.

C. K. Rofer–De Poorter, 'A Comprehensive Mechanism for the Fischer–Tropsch Synthesis', *Chem. Rev.*, 81, 447 (1981).

CHAPTER 4

OLEFIN HYDROFORMYLATION

Unsaturated hydrocarbons, especially C_2-C_4 olefins, are important building blocks in the petrochemical industry. They are converted on a large scale to a variety of oxygenated derivatives such as alcohols, aldehydes, ketones and carboxylic acids. The introduction of oxygen into the molecule generally involves reaction of the olefin with one of the three abundantly available reagents: water, molecular oxygen or carbon monoxide. Examples of reactions of olefins with water and molecular oxygen are the conversion of propylene to isopropanol (Reaction 1) and acrolein (Reaction 2), respectively.

$$CH_3CH{=}CH_2 + H_2O \xrightarrow{\;H^+\;} CH_3CH(OH)CH_3 \tag{1}$$

$$CH_3CH{=}CH_2 + \tfrac{1}{2}O_2 \xrightarrow{[Bi_2MoO_4]} CH_2{=}CHCHO \tag{2}$$

The third possibility, reaction with carbon monoxide, forms the subject of this and the following chapter. At the present time several million tons of industrial chemicals are produced annually via the catalytic reactions of carbon monoxide with olefins. The most important and oldest of these reactions is olefin hydroformylation.

Hydroformylation is the general term applied to the reaction of an olefin with a mixture of carbon monoxide and hydrogen (syn gas) to form an aldehyde. The reaction occurs only in the presence of certain transition metal carbonyl catalysts, usually of cobalt or rhodium. The simplest example of the reaction is the hydroformylation of ethylene to propionaldehyde.

$$CH_2{=}CH_2 + CO + H_2 \longrightarrow CH_3CH_2CHO \tag{3}$$

When a substituted, terminal olefin is used both linear and branched aldehydes are formed. Propylene, for example, affords a mixture of n-butyraldehyde and isobutyraldehyde (Reaction 4). In general the linear aldehyde is the major product.

$$CH_3CH{=}CH_2 + CO + H_2 \longrightarrow CH_3CH_2CH_2CHO + (CH_3)_2CHCHO \tag{4}$$

The reaction was discovered by Roelen in 1938 [1], while investigating the effect of added olefins on the Fischer–Tropsch process in the presence of a heterogeneous cobalt catalyst. Later it was shown that the actual catalyst is a soluble cobalt carbonyl species with the reaction taking place in a homogeneous liquid phase [2–4]. Indeed, hydroformylation is presently the most important industrial application of homogeneous, transition metal catalysis. The reaction is often referred to as the Oxo process but the more correct term is hydroformylation, first coined by Adkins [5].

Because of its commercial importance the hydroformylation reaction has been studied extensively and the subject has been frequently reviewed [6–12].

4.1 Commercial Application

The most important industrial application of hydroformylation is the manufacture of n-butyraldehyde from propylene and syn gas (Reaction 4). The aldehyde is an important chemical intermediate and is produced on a scale of about 3 million tons per year worldwide. Some of the butyraldehyde is hydrogenated to n-butanol which is used extensively as an industrial solvent. The major part of the n-butyraldehyde is converted to 2-ethylhexanol via aldol condensation and hydrogenation (Scheme 4.1). Much of the 2-ethylhexanol is converted to phthalate esters which are used as plasticizers in polyvinyl chloride (PVC).

$$2 \ CH_3CH_2CH_2CHO \xrightarrow[-H_2O]{base} CH_3(CH_2)_2CH=C(C_2H_5)CHO$$

$$\xrightarrow{H_2} CH_3(CH_2)_3CH(C_2H_5)CH_2OH$$

Scheme 4.1

A second major application of hydroformylation is in the synthesis of long chain alcohols from terminal olefins such as 1-octene and higher. The process sometimes combines hydroformylation and hydrogenation steps since phosphine modified cobalt carbonyl catalysts (see later) are effective for both reactions:

$$RCH=CH_2 \xrightarrow[H_2]{CO} RCH_2CH_2CHO \xrightarrow{H_2} R(CH_2)_3OH \qquad (5)$$

The linear, fatty alcohols so produced, particularly those in the C_{13}–C_{15} range, are used in the manufacture of biodegradable detergents.

The hydroformylation of ethylene to propionaldehyde (Reaction 3) constitutes the third major industrial application of hydroformylation. Propionaldehyde has no important uses as such but is converted to n-propanol and propionic acid by hydrogenation and oxidation, respectively.

The majority of industrial hydroformylation processes employ cobalt catalysts and require temperatures of 120–140°C and pressures of $ca.$ 200 bar. The high pressure is necessary in order to stabilize the cobalt carbonyl intermediates and the high temperature reflects the conditions needed to generate the active catalyst, $HCo(CO)_4$. When the latter is generated in advance in a separate reactor the hydroformylation proceeds smoothly at 90–120°C. The need for milder reaction conditions and higher product selectivity (linear to branched ratio) stimulated extensive research in hydroformylation technology. This led to the development of the more selective phosphine-modified cobalt catalysts and the considerably more active rhodium-based catalysts (see below).

In addition to the petrochemical applications outlined above, hydroformylation constitutes a useful method for the synthesis of aldehydes from a variety of terminal and internal olefins. Such syntheses can be carried out at atmospheric pressure by using $HCo(CO)_4$ in stoichiometric quantities or by using the much more active rhodium-based catalysts such as $HRh(CO)(Ph_3P)_3$. Thus, whilst the commercial application of hydroformylation has, up till now, been almost exclusively in the bulk chemical area there is every reason to believe that the reaction will be applied successfully to the production of fine organic chemicals in the future.

4.2 Mechanism

Subsequent to the pioneering work of Roelen [1] it was recognized that the hydroformylation reaction is catalysed by a soluble species and $HCo(CO)_4$ was proposed to be the active catalyst [2–5]. This was substantiated by studies of the stoichiometric hydroformylation of olefins with $HCo(CO)_4$ [3]. Typically, the catalyst is generated by treating finely divided cobalt or a cobalt(II) salt with syn gas at elevated temperatures and pressures. This results in the formation of $HCo(CO)_4$ via the following sequence of transformations (Scheme 4.2).

$$Co^{II} \xrightarrow[CO]{H_2} Co \xrightarrow{CO} Co_2(CO)_8 \xrightarrow{H_2} HCo(CO)_4$$

Scheme 4.2

For the subsequent reaction of $HCo(CO)_4$ with the olefin the mechanism proposed by Heck and Breslow [13, 14] is the one most widely accepted. This mechanism is outlined in Scheme 4.3 for ethylene as substrate.

$$HCo(CO)_4 \rightleftharpoons HCo(CO)_3 + CO \qquad (6)$$

$$HCo(CO)_3 + CH_2{=}CH_2 \rightleftharpoons HCo(CO)_3 \rightleftharpoons CH_3CH_2Co(CO)_3 \qquad (7)$$
$$\uparrow$$
$$CH_2{=}CH_2$$

$$CH_3CH_2Co(CO)_3 \underset{\xleftarrow{\hspace{1cm}}}{\overset{CO}{\xrightleftharpoons{\hspace{1cm}}}} CH_3CH_2Co(CO)_4 \tag{8}$$

$$\xrightleftharpoons{\hspace{1cm}} CH_3CH_2COCo(CO)_3$$

$$\text{(I)}$$

$$I \xrightarrow{\hspace{1cm}} \begin{cases} \xrightarrow{H_2} CH_3CH_2\underset{\overset{|}{H}}{\overset{\overset{H}{|}}{COCo(CO)_3}} \longrightarrow CH_3CH_2CHO + HCo(CO)_3 \tag{9} \\ \\ \xrightarrow{HCo(CO)_4} CH_3CH_2CHO + Co_2(CO)_7 \tag{10} \\ \qquad\qquad\qquad\qquad\qquad \Large\lfloor \longrightarrow Co_2(CO)_8 \end{cases}$$

Scheme 4.3

When the olefin is substituted, i.e. $RCH=CH_2$, insertion of the olefin into the Co–H bond results in the formation of two isomeric alkylcobalt intermediates, $RCH_2CH_2Co(CO)_3$ and $RCH(CH_3)Co(CO)_3$. Subsequent CO insertion and hydrogenolysis produces the isomeric linear and branched aldehydes, RCH_2CH_2CHO and $RCH(CH_3)CHO$, respectively.

According to Orchin [15] the product distribution (linear vs. branched) is determined by the relative rates of CO insertion into the isomeric alkylcobalt intermediates. These relative rates are in turn dependent on the presence or absence of nucleophiles that can coordinate to the cobalt. Thus, at high partial pressures of CO or in the presence of added nucleophiles, the acyl species (I) is probably five-coordinate, with its formation via CO migratory insertion likely involving a six-coordinate transition state. In such a situation, the greater steric requirement of the transition state for the branched isomer increases the energy of this transition state relative to that of the linear isomer. The respective transition states for the linear and branched isomers are depicted in Figure 4.1.

Figure 4.1. Transition states for migratory CO insertion.

The overall result is that the linear isomer should be favoured at high partial pressures of CO and in the presence of added nucleophiles. This is consistent with the well-known enhancement of linear to branched aldehyde ratios observed with phosphine-modified cobalt catalysts such as $Co_2(CO)_6(Bu_3P)_2$ (see below).

Alternatively, the different linear-to-branched ratios may be due to the influence of phosphine ligands on the polarity of the Co–H bond, perhaps even resulting in a different mode of addition of the cobalt carbonyl hydride intermediate to the olefin.

In Scheme 4.3 two possible pathways are depicted for the hydrogenolysis of the key acylcobalt intermediate (I). Heck and Breslow [13, 14] favoured a pathway involving the oxidative addition of hydrogen followed by reductive elimination of the aldehyde (Reaction 9). On the other hand, in the stoichiometric hydroformylation of olefins (I) must be converted to aldehydes by reaction with $HCo(CO)_4$ (Reaction 10) since this is the sole source of the aldehydic hydrogen. Indeed, recent IR studies under typical reaction conditions indicate that Reaction 10 is the predominant pathway even in the catalytic hydroformylation reaction [16]. Reaction 10 may be envisaged as a direct protonolysis of (I) by the strongly acidic $HCo(CO)_4$.

Hydroformylation of straight chain olefins containing the double bond in the two position affords predominantly the linear aldehyde, the product expected from the terminal olefin. For example, the catalytic hydroformylation of 2-pentene yields almost as much 1-hexanal as does 1-pentene under identical conditions [17], i.e.,

$$CH_3CH_2CH{=}CHCH_3 \xrightarrow[H_2]{CO} CH_3(CH_2)_4CHO \xleftarrow[H_2]{CO} CH_3CH_2CH_2CH{=}CH_2$$

major product

Significantly, little or no 1-pentene is formed during the course of the reaction. Detailed mechanistic studies with chiral [18] and deuterium-labelled [19] internal olefins demonstrated that the formation of such isomeric aldehydes involves the non-dissociative migration of the cobalt along the carbon chain. This is illustrated in Scheme 4.4 for the 2-pentene to 1-hexanal conversion.

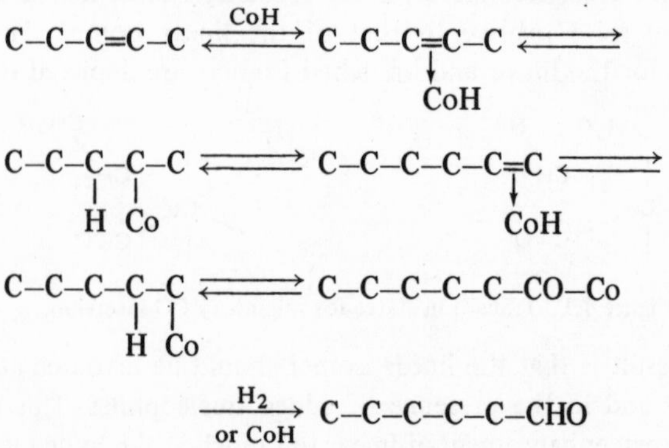

$$C{-}C{-}C{=}C{-}C \underset{}{\overset{CoH}{\rightleftharpoons}} C{-}C{-}C{=}C{-}C \rightleftharpoons$$
$$\underset{\substack{\downarrow \\ CoH}}{}$$

$$C{-}C\underset{\substack{| \\ H}}{-}C\underset{\substack{| \\ Co}}{-}C{-}C \rightleftharpoons C{-}C{-}C{-}C{-}C{=}C \rightleftharpoons$$
$$\underset{\substack{\downarrow \\ CoH}}{}$$

$$C{-}C{-}C\underset{\substack{| \\ H}}{-}C\underset{\substack{| \\ Co}}{-}C \rightleftharpoons C{-}C{-}C{-}C{-}C{-}CO{-}Co$$

$$\xrightarrow[\text{or CoH}]{H_2} C{-}C{-}C{-}C{-}C{-}CHO$$

Scheme 4.4

Rhodium-catalysed hydroformylations probably follow a similar pathway to that outlined in Scheme 4.3 although in this case the reaction is further complicated by the existence of equilibria (Scheme 4.5) between the putative intermediate $HRh(CO)_3$ and rhodium clusters [20]. The formation of multinuclear clusters is much more favourable for rhodium than for cobalt.

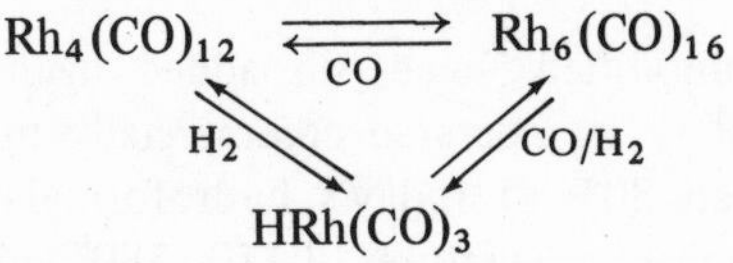

$$Rh_4(CO)_{12} \; \underset{CO}{\overset{}{\rightleftarrows}} \; Rh_6(CO)_{16}$$

$$H_2 \searrow \qquad \swarrow CO/H_2$$

$$HRh(CO)_3$$

Scheme 4.5

Although the reaction steps as outlined for cobalt in Scheme 4.3 are substantially the same in rhodium-catalysed hydroformylations, the relative rates of the individual steps appear to be significantly different for the two catalysts [21]. Indeed the rate-limiting step in Scheme 4.3 varies not only with the catalyst used but also with the reactivity of the olefin substrate and the CO partial pressure [21].

In the generally accepted mechanism as discussed above the individual steps all involve heterolytic processes and 16 and 18 electron intermediates. As we have noted in Chapter 2, however, there is increasing evidence that reactions of metal carbonyls and carbonyl hydrides can also involve homolytic processes and paramagnetic species as obligatory intermediates. It is worth noting, therefore, that homolytic mechanisms can be envisaged for individual steps in Scheme 4.3. For example, the following free radical chain propagation sequence constitutes a plausible alternative to direct protonolysis for Reaction 10.

$$R\overset{\cdot}{C}O + \overset{\cdot}{H}Co(CO)_4 \longrightarrow RCHO + Co(CO)_4 \qquad (11)$$

$$Co(CO)_4 + RCOCo(CO)_3 \longrightarrow Co_2(CO)_7 + R\overset{\cdot}{C}O \qquad (12)$$

Further work is obviously necessary to distinguish unequivocally between homolytic and heterolytic mechanisms for many of these reactions.

4.3 Catalysts and Processes

A comparison of the hydroformylation activity of various transition metals indicates a decided preference for cobalt and rhodium [22]:

$$Rh > Co > Ru \; > Mn \; > Fe \; > Cr, Mo, W, Ni$$

$$10^3-10^4 \quad 1 \quad 10^{-2} \quad 10^{-4} \quad 10^{-6} \quad\quad 0$$

Three types of homogeneous transition metal catalysts are employed in industrial hydroformylation processes. In order of their historical development and present commercial importance these are:

(a) simple cobalt carbonyls
(b) phosphine-modified cobalt carbonyls
(c) phosphine-modified rhodium carbonyls

Processes using unmodified — i.e. no added ligands — cobalt carbonyl catalysts were the first to be operated commercially by Ruhrchemie and still account for more than 80% of today's hydroformylation capacity. Typical reaction conditions involve temperatures of $110-180°$ and pressures in the range $200-350$ bar. The simple cobalt carbonyl catalysts ($HCo(CO)_4$ is the actual catalyst) are reasonably satisfactory for industrial hydroformylation processes, but several problems do exist. The high volatility and low stability of $HCo(CO)_4$ and $Co_2(CO)_8$ makes separation from the aldehyde products cumbersome. More serious limitations are the severe reaction conditions required and the moderate selectivities to the desired linear aldehydes. Thus, typical linear to branched aldehyde ratios observed are $3-4 : 1$, i.e. ca. 75–80% linear aldehyde selectivity.

Because of the limitations of the simple cobalt catalysts much research effort was devoted to developing improved catalysts. This led to the introduction of the phosphine-modified cobalt and rhodium catalysts.

Compared to the unmodified systems, cobalt carbonyl catalysts containing added tertiary phosphine ligands, e.g. Bu_3P, are considerably more stable and pressures as low as $5-10$ bar can be used at temperatures in the range $100-180°C$ [23, 24]. In this process, developed by Shell workers [23, 24], the active catalyst appears to be $HCo(CO)_3(R_3P)$. These phosphine-modified catalysts are advantageous for the direct conversion of terminal olefins to linear alcohols and the Shell process is applied mainly to the manufacture of fatty alcohols from C_8-C_{15} alpha olefins.

$$RCH{=}CH_2 + CO + 2H_2 \xrightarrow{[HCo(CO)_3(R_3P)]} RCH_2CH_2CH_2OH \qquad (13)$$

$$R = n\text{-}C_6-C_{13}$$

When simple cobalt carbonyls are used as hydroformylation catalysts the aldehyde is hydrogenated in a separate step using a conventional hydrogenation catalyst. The phosphine-modified cobalt carbonyls, in contrast, are efficient hydrogenation catalysts and the hydroformylation and hydrogenation can be carried out in a single stage as indicated in Equation 13. A further advantage is the higher linear : branched ratio compared to that observed with unmodified catalysts (ca. 8 : 1 and 4 : 1, respectively). The increased selectivity to linear products is thought to be due to steric effects of the bulky phosphine ligand in the transition state (see preceding section).

The stability of the modified cobalt catalysts simplifies their recovery and recycle since the alcohol product can be distilled from the catalyst. A drawback of the process is the loss of olefin (*ca.* 10–15%) through competing hydrogenation to the corresponding alkane. Another limitation is the lower activity of the modified catalyst; hydroformylation with $HCo(CO)_4$ at 145°C is *ca.* 5 times faster than with $HCo(CO)_3R_3P$ at 180°C.

The superior activity of rhodium carbonyls (10^3–10^4 times as active as cobalt carbonyls) was demonstrated in industrial laboratories in the 1950s. Their commercial application was hampered by the fact that they afford poor linear/branched aldehyde ratios. This problem was solved with the discovery that phosphine-modified rhodium catalysts exhibit both high activity and high selectivity to linear aldehydes [25, 26]. This modified rhodium hydroformylation was commercially implemented by Union Carbide [27, 28].

The commercial catalyst is prepared by reaction of metallic rhodium with syn gas in the presence of triphenylphosphine or a phosphite. The immediate catalyst precursor is $HRh(CO)(Ph_3P)_3$. This complex catalyses the hydroformylation of olefins readily at atmospheric pressure, but commercial operation generally involves pressures of 10–20 bar and a temperature of *ca.* 100°C [27–29]. The selectivity to linear aldehydes is high, being greater than 90% in the presence of excess phosphine ligand.

The mechanism of hydroformylation with the modified rhodium catalysts is essentially the same as that with cobalt carbonyl catalysts [30, 31]. As in the cobalt system the initial step is the generation of a reactive, coordinatively unsaturated species by ligand dissociation (Reaction 14).

$$HRh(CO)(Ph_3P)_3 \rightleftharpoons HRh(CO)(Ph_3P)_2 + Ph_3P \qquad (14)$$

This is followed by the sequence of reactions outlined in Scheme 4.6 with ethylene as the substrate (compare Scheme 4.3).

$$HRh(CO)L_2 \xrightleftharpoons{CH_2=CH_2} HRh(CO)L_2 \rightleftharpoons CH_3CH_2Rh(CO)L_2$$

$$\uparrow CH_2=CH_2$$

$$\xrightleftharpoons{CO} CH_3CH_2Rh(CO)_2L_2 \rightleftharpoons CH_3CH_2CORh(CO)L_2$$

$$\xrightarrow{H_2} CH_3CH_2CORhL_2 \xrightarrow{\quad} CH_3CH_2CHO + HRh(CO)L_2$$

with H above and H below on the CORhL₂ species.

$L=Ph_3P$

Scheme 4.6

A serious drawback of the rhodium-based catalysts is, of course, the high cost of rhodium, which is more expensive than gold. This requires that catalyst losses be kept to an absolute minimum as even losses in the ppm range can have a significant effect on the process economics. In the Union Carbide process for the hydroformylation of propylene this problem is solved by using a gas-sparged reactor in which the mixture of propylene and syn gas are sparged through the catalyst solution [32]. The butyraldehyde product is continuously removed in the gaseous effluent and catalyst losses are negligible since the catalyst never leaves the reactor.

Another approach to the problem of catalyst recovery and recycle is to immobilize it by chemically bonding the transition metal complex to an insoluble support. The reaction can then, in principle, be carried out in fixed bed operation whilst retaining the high activity and selectivity characteristic of homogeneous catalysts. The most widely studied supports for this purpose are silica and alumina [33, 34] and cross-linked polystyrene resins containing pendant phosphine ligands [34–37]. Examples of reactions used to bind metal complexes to such supports are illustrated in Figure 4.2.

As with conventional heterogeneous catalysts, diffusion of reactants and products within a catalyst is often a rate-limiting step. With the polystyrene-

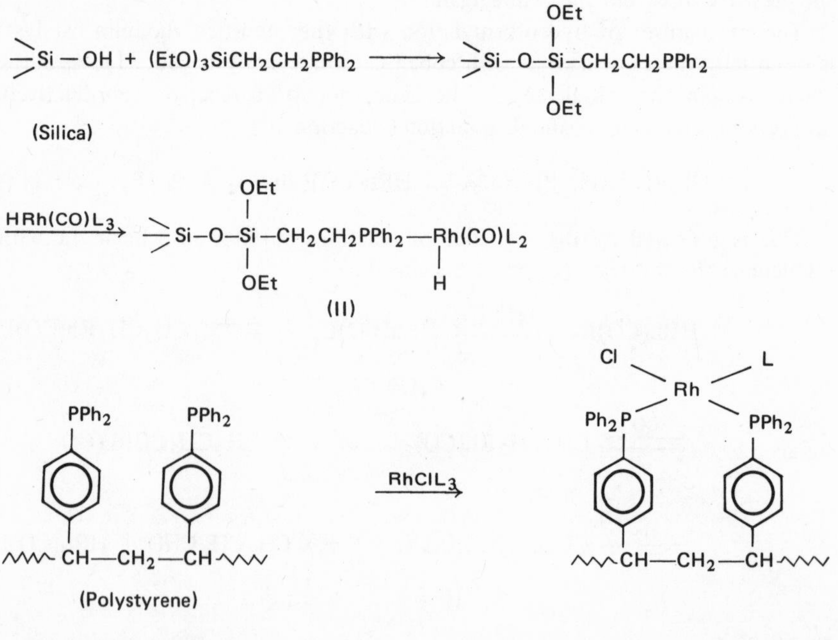

Figure 4.2. Immobilization of rhodium complexes on silica and polystyrene supports.

based catalysts extensive swelling often occurs in organic media due to diffusion of the solvent into the polymer lattice. This can be advantageous from the point of view of increasing the rate of diffusion of reactants and products but can lead to disruption of the catalyst particle.

With oxide supports, on the other hand, swelling is not observed and diffusion occurs through the pore structure of the support. For this reason silica and alumina supports have received more attention in industrial laboratories. Silica-supported rhodium catalysts such as (II) have been shown to be particularly effective hydroformylation catalysts [38, 39]. These catalysts have, as yet, not been commercialized but they are likely to find a place in future industrial practice. It is worth noting, that these catalysts can be conveniently deployed in small-scale hydroformylations where their high activity and ease of separation are particularly advantageous.

Summarising, the major characteristics of the three types of commercial catalysts are collected in Table 4.I.

TABLE 4.I
Comparison of hydroformylation catalysts

	$HCo(CO)_4$	$HCo(CO)_3(Bu_3P)$	$HRh(CO)(Ph_3P)_3$
Temperature (°C)	140−180	160−200	60−120
Pressure (bar)	200−350	50−100	1−50
Catalyst concn. (% metal/olefin)	0.1−1.0	0.5−1.0	$10^{-2}-10^{-3}$
Hydrocarbon formation (%)	*ca.* 1	10−15	2−5
Reaction products	aldehydes	alcohols	aldehydes
Linear/branched ratio	80−20	*ca.* 88 : 12	up to 92 : 8

The rhodium-based catalysts, because of their superior activity, are likely to enjoy increasing commercial application in the future as the contribution of energy costs to process economics becomes increasingly important. Finally, active and selective hydroformylation catalysts are reportedly formed when $PtCl_2L_2$ (L= tertiary phosphine) is reacted with $SnCl_2$ [40]. The active catalyst is thought to be the hydridoplatinum(II) complex, $HPt(SnCl_3)(CO)(R_3P)$, which is some 5 times as active as a modified cobalt catalyst [41]. This type of catalyst has not yet been commercialized.

4.4 Olefin Substrates

The relative rates of hydroformylation of various types of olefins follow much the same pattern with both cobalt [42] and rhodium [43] catalysts. Some typical rate constants for $Co_2(CO)_8$-catalysed hydroformylations are collected in Table 4.II. The relative rates decrease in the order:

$$CH_2{=}CH_2 > RCH{=}CH_2 > RCH{=}CHR'$$
$$> RR'C{=}CH_2 > RR'C{=}CHR'' > RR'C{=}CR''R'''$$

TABLE 4.II

Rates of $Co_2(CO)_8$-catalysed hydroformylation of various olefins[a]

Olefin	$k \times 10^3$ (min^{-1})
Straight-chain terminal	
1-Hexene	66.2
1-Octene	66.8
1-Decene	64.4
1-Tetradecene	63.0
Straight-chain internal	
2-Hexene	18.1
2-Heptene	19.3
3-Heptene	20.0
Branched terminal	
4-Methyl-1-pentene	64.3
2-Methyl-1-pentene	7.3
2,4,4-Trimethyl-1-pentene	4.8
Branched internal	
2-Methyl-2-pentene	4.9
2,3-Dimethyl-2-butene	1.4
Cyclic	
Cyclopentene	22.4
Cyclohexene	5.8
Cycloheptene	25.7
Cyclooctene	10.8

[a] 0.5 mole olefin + 0.008 mole $Co_2(CO)_8$ in 65 mL methylcyclohexane at $110°$ and 233 bar ($CO/H_2{=}1/1$). Data taken from Wender *et al.* [42].

The position of the double bond in a linear, internal olefin has little effect on the rate, e.g. 2-heptene and 3-heptene react equally fast. Cyclic olefins exhibit a rather irregular pattern, but all are less reactive than terminal olefins (see Table 4.II). Branching at an olefinic carbon atom has a retarding effect on the rate and hydroformylation is generally not observed at a quaternary carbon [44]. Thus, hydroformylation of 2,3-dimethyl-2-butene is slow and affords 3,4-dimethylpentanal via initial isomerization.

$$(CH_3)_2C{=}C(CH_3)_2 + CO + H_2$$
$$\longrightarrow (CH_3)_2CHCH(CH_3)CH_2CHO \qquad (15)$$

In addition to being an important petrochemical process hydroformylation constitutes a useful synthetic method in organic chemistry. For example, the rhodium-catalysed hydroformylation of the diene (III) to aldehyde (IV) (Reaction 16) is a key step in an isoprene-based route to the Vitamin E intermediate, phytone [45].

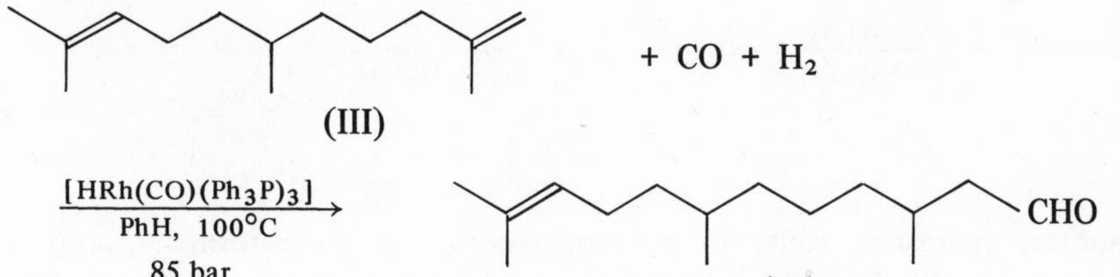

$$+ \ CO \ + \ H_2$$

(III)

$$\xrightarrow[\substack{PhH, \ 100°C \\ 85 \ bar}]{[HRh(CO)(Ph_3P)_3]}$$

CHO   (16)

(IV)

Similarly, the hydroxyaldehyde (VI), of interest as an aroma chemical, was prepared by rhodium-catalysed hydroformylation of (V) in benzene at 100°C and 85 bar [46].

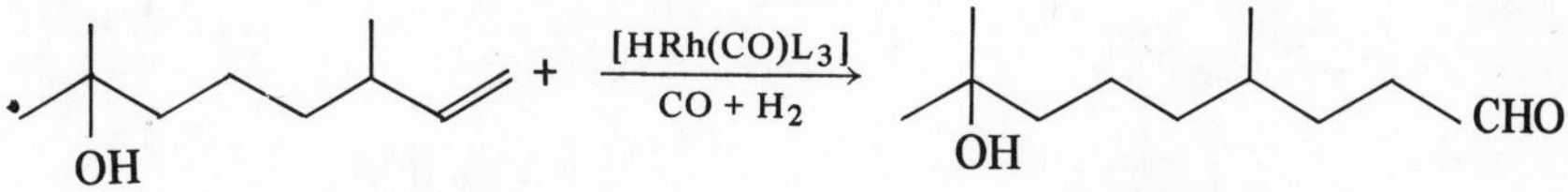

$$\xrightarrow[\substack{CO + H_2}]{[HRh(CO)L_3]}$$

CHO

OH

(V)

OH

(VI)

(17)

A reaction of potential industrial significance is the hydroformylation of allyl alcohol to give, after hydrogenation (see Scheme 4.7), 1,4-butanediol [47, 48], which is a raw material for the manufacture of high performance engineering plastics.

$$CH_2{=}CHCH_2OH \ + \ CO \ + \ H_2 \ \xrightarrow[\substack{30°C, \ 1-2 \ bar}]{[HRh(CO)(Ph_3P)_3]}$$

$$HO(CH_2)_3CHO$$
$$+$$
$$HOCH_2CH(CH_3)CHO$$

$$\xrightarrow{H_2}$$

$$HO(CH_2)_4OH \qquad (77\%)$$
$$+$$
$$HOCH_2CH(CH_3)CH_2OH \qquad (20\%)$$

Scheme 4.7

1,4-Butanediol is currently manufactured by reaction of formaldehyde with acetylene followed by hydrogenation:

$$HC{\equiv}CH + 2\,H_2CO \longrightarrow HOCH_2C{\equiv}CCH_2OH \xrightarrow{H_2} HO(CH_2)_4OH \quad (18)$$

Allyl alcohol is available from propylene via, for example, catalytic oxidation to allyl acetate and subsequent hydrolysis or via isomerization of propylene oxide (Scheme 4.8).

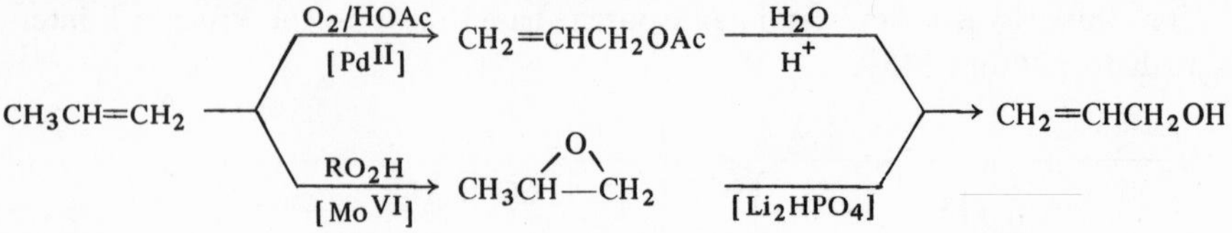

Scheme 4.8

Another, interesting route to 1,4-butanediol is via hydroformylation of acrolein acetals [49–51] whereby the unwanted isomeric diol is used for acetal formation (see Scheme 4.9). Acrolein is readily available from the catalytic oxidation of propylene over bismuth molybdate (see Chapter 1).

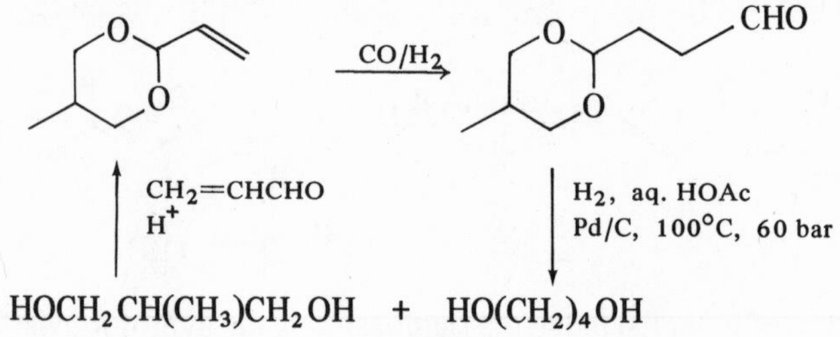

$$HOCH_2CH(CH_3)CH_2OH \quad + \quad HO(CH_2)_4OH$$

Scheme 4.9

4.5 Asymmetric Hydroformylation

When the hydroformylation reaction is applied to a prochiral olefin a chiral centre is generated (Reaction 19). When a catalyst containing optically active (chiral) ligands is used this offers the possibility of asymmetric induction, i.e. catalytic asymmetric hydroformylation [52–54].

$$RR'C{=}CH_2 + CO + H_2 \longrightarrow RR'CHCH_2CHO \qquad (19)$$

The best results have been obtained with rhodium catalysts containing the chiral phosphine ligands DIOP and DIPHOL (see below). However, the observed asymmetric inductions have generally been poor to moderate, e.g. up to 27% and 44% enantiomeric excess (ee) with DIOP [52, 53] and DIPHOL [54], respectively.

Although the published results have generally been disappointing it was pointed out [55] that only a limited selection of the arsenal of available chiral phosphine ligands has been tried and mainly with non-functionalized olefins as substrates. Experience with the related asymmetric hydrogenation reaction has

DIOP   DIPHOL

shown that high optical yields are generally observed only with olefins containing polar substituents such as dehydroamino acids [56]. It is not encouraging, however, that the asymmetric hydroformylation of enamides only gave poor to moderate optical yields (20–40% ee) using DIOP and DIPHOL as ligands [57].

4.6 Hydroformylation with CO/H$_2$O

Olefin hydroformylation can also be accomplished using a mixture of carbon monoxide and water, in basic media, instead of syn gas [58, 59]. Fe(CO)$_5$ is the preferred catalyst for these reactions which lead to the formation of alcohols (Reaction 20). The reaction is generally referred to as the Reppe alcohol synthesis after its discoverer [59].

$$RCH{=}CH_2 \ + \ 3\,CO \ + \ 2\,H_2O \ \longrightarrow \ R(CH_2)_3OH \ + \ 2\,CO_2 \qquad (20)$$

The reaction is commercially less attractive than conventional hydroformylation because of the much higher consumption of carbon monoxide. It is, however, used commercially in the BASF process for the manufacture of n-butanol from propylene (Reaction 21). In this process n-butanol and isobutanol are produced in 85% and 15% yield, respectively, by reaction of CO/H$_2$O with propylene at 100°C and 15 bar in the presence of Fe(CO)$_5$ and N-propylpyrrolidine as base [60].

$$CH_3CH{=}CH_2 \ + \ 3\,CO \ + \ 2\,H_2O \ \longrightarrow \ CH_3(CH_2)_3OH \ + \ 2\,CO_2 \qquad (21)$$

The actual catalyst in these reactions is H$_2$Fe(CO)$_4$ formed via Reaction 22, which is the key step in the water gas shift reaction.

$$Fe(CO)_5 + H_2O \xrightarrow{\text{base}} H_2Fe(CO)_4 + CO_2 \qquad (22)$$

$Fe(CO)_5$ is a poor catalyst in conventional hydroformylations with CO/H_2 because the formation of $H_2Fe(CO)_4$ by reaction of $Fe(CO)_5$ with H_2 is unfavourable. The formation of $H_2Fe(CO)_4$ via Reaction 22, in contrast, proceeds under relatively mild conditions. The subsequent reaction of $H_2Fe(CO)_4$ with the olefin involves a conventional hydroformylation mechanism as shown in Scheme 4.10 (see also Section 2.7).

$$H_2Fe(CO)_4 + RCH=CH_2 \longrightarrow RCH_2CH_2Fe(CO)_4H$$

$$\xrightarrow{CO} RCH_2CH_2COFe(CO)_4H$$

$$\xrightarrow{H_2} H_2Fe(CO)_4 + RCH_2CH_2CHO \xrightarrow{H_2Fe(CO)_4} R(CH_2)_3OH$$

Scheme 4.10

4.7 Hydroformamination

When the hydroformylation reaction is carried out in the presence of secondary amines this results in the formation of tertiary amines according to the following stoichiometry:

$$RCH=CH_2 + CO + 2H_2 + R_2NH \longrightarrow R(CH_2)_3NR_2 + H_2O \qquad (23)$$

The reaction occurs under roughly the same conditions and with the same catalysts as conventional hydroformylation. It probably involves the intermediate formation of an enamine via condensation of the initially formed aldehyde with the secondary amine (Scheme 4.11).

$$RCH=CH_2 + CO + H_2 \longrightarrow RCH_2CH_2CHO$$

$$\xrightarrow[-H_2O]{R_2NH} RCH_2CH=CHNR_2 \xrightarrow{H_2} R(CH_2)_3NR_2$$

Scheme 4.11

The reaction is referred to as hydroformamination or aminomethylation [58] and can also be carried out using CO/H_2O mixtures instead of syn gas. In this case good results have been reported using mixed metal carbonyl catalysts such as $Rh_6(CO)_{16}/Fe_3(CO)_{12}$ and $Ru_3(CO)_{12}/Fe_3(CO)_{12}$ [58]. Tertiary amines are formed according to the stoichiometry:

$$RCH=CH_2 + 3CO + H_2O + R_2NH \longrightarrow R(CH_2)_3NR_2 + 2CO_2 \qquad (24)$$

4.8 Summary

Olefin hydroformylation is a reaction of considerable commercial importance. It has broad synthetic utility as a method for converting olefins to industrially important derivatives such as aldehydes, alcohols, amines and carboxylic acids (Scheme 4.12).

$$RCH{=}CH_2 \quad
\begin{array}{l}
\xrightarrow{\;CO/H_2\;} R(CH_2)_3 OH \\[2pt]
\qquad\qquad\quad \uparrow H_2 \\[2pt]
\xrightarrow{\;CO/H_2\;} RCH_2 CH_2 CHO \xrightarrow{\;O_2\;} RCH_2 CH_2 CO_2 H \\[2pt]
\xrightarrow[R_2NH]{\;CO/H_2\;} R(CH_2)_3 NR_2
\end{array}$$

Scheme 4.12

Although the reaction has, up till now, mainly been applied to the industrial manufacture of bulk chemicals once organic chemists fully recognize its potential it could become a widely used synthetic technique in organic chemistry. Further variants of the reaction also offer considerable promise in this respect. For example, the hydroformylation of olefins in the presence of amides provides a route to amino acid derivatives (see Chapter 8).

References

1. O. Roelen, *German Patent* 849,548 (1938) to Ruhr Chemie.
2. I. Wender, M. Orchin and H. H. Storch, *J. Am. Chem. Soc.*, **72**, 4842 (1950).
3. I. Wender, H. W. Sternberg and M. Orchin, *J. Am. Chem. Soc.*, **75**, 3041 (1953).
4. L. Kirch, I. J. Goldfarb and M. Orchin, *J. Am. Chem. Soc.*, **78**, 5450 (1956).
5. H. Adkins and G. Krsek, *J. Am. Chem. Soc.*, **71**, 3051 (1949).
6. M. Orchin and W. Rupilius, *Catal. Rev.*, **6**, 85 (1972).
7. F. E. Paulik, *Catal. Rev.*, **6**, 49 (1972).
8. A. J. Chalk and J. F. Harrod, *Advan. Organometal. Chem.*, **11**, 119 (1968).
9. L. Marko, in *Aspects of Homogeneous Catalysis* (R. Ugo, Ed.) Vol. 2, Chapter 1, Reidel, Dordrecht, 1973.
10. P. Pino, F. Piacenti and M. Bianchi, in *Organic Synthesis via Metal Carbonyls* (I. Wender and P. Pino, Eds.) Vol. 2, Wiley, New York, 1977, p. 43.
11. R. L. Pruett, *Advan. Organometal. Chem.*, **17**, 1 (1974).
12. B. Cornils in *New Syntheses with Carbon Monoxide*, (J. Falbe, Ed.), Springer-Verlag, Berlin, 1980, p. 1.
13. R. F. Heck and D. S. Breslow, *J. Am. Chem. Soc.*, **83**, 4023 (1961).
14. D. S. Breslow and R. F. Heck, *Chem. Ind.* (London), p. 467 (1960).
15. M. Orchin, *Acc. Chem. Res.*, **14**, 259 (1981).
16. N. H. Alemdarogly, J. L. M. Penninger and E. Oltay, *Monatsch Chem.*, **107**, 1153 (1976).

17. I. J. Goldfarb and M. Orchin, *Advan. Catal.*, **14**, 1 (1957).
18. F. Piacenti, S. Pucci, M. Bianchi, R. Lazzaroni and P. Pino, *J. Am. Chem. Soc.*, **90**, 6847 (1968).
19. C. P. Casey and C. R. Cyr, *J. Am. Chem. Soc.*, **93**, 1280 (1971); **95**, 2240 (1973).
20. G. Csontos, B. Heil and L. Marko, *Ann. N. Y. Acad. Sci.*, **239**, 47 (1974).
21. P. Pino, *J. Organometal. Chem.*, **200**, 223 (1980).
22. K. A. Alekseeva, M. D. Vysotski, N. S. Imyanitov and V. A. Rybakov, *Zh. Vses. Khim. O-va.*, **22**, 45 (1977); *C. A.* **86**, 120712f (1977).
23. L. H. Slaugh and R. D. Mullineaux, *US Patents* 3,239,569 and 3,239,570 (1966) to Shell.
24. L. H. Slaugh and R. D. Mullineaux, *J. Organometal. Chem.*, **13**, 469 (1968).
25. L. H. Slaugh and R. D. Mullineaux, *US Patent* 3,239,566 (1966) to Shell.
26. J. A. Osborn, J. F. Young and G. Wilkinson, *J. Chem. Soc. Chem. Commun.*, 17 (1965).
27. R. Fowler, H. Connor and R. A. Baehl, *Hydrocarbon Process.*, **55** (9), 247 (1976); *Chem. Tech.*, 772 (1976).
28. R. L. Pruett, *Ann. N. Y. Acad. Sci.*, **295**, 239 (1977).
29. B. Cornils, R. Payer and K. C. Traenckner, *Hydrocarbon Process.*, **54** (6), 83 (1975).
30. D. Evans, G. Yagupsky and G. Wilkinson, *J. Chem. Soc. A*, 2660 (1968).
31. G. Yagupsky, C. K. Brown and G. Wilkinson, *J. Chem. Soc. A*, 1392, 2753 (1970).
32. A. Hershman, K. K. Robinson, J. H. Craddock and J. F. Roth, *Ind. Eng. Chem. Prod. Res. Dev.*, **8**, 372 (1969).
33. L. L. Murrell, in *Advanced Materials in Catalysis* (J. J. Burton and R. L. Garten, Eds.) Academic Press, New York, 1977, p. 235.
34. R. H. Grubbs, *Chem. Tech.*, 512 (1977).
35. R. H. Grubbs, in *Enzymic and Non-Enzymic Catalysis*, (P. Dunnill, A. Wiseman and N. Blakebrough, Eds.), Ellis Horwood, Chichester, 1980, p. 224.
36. Z. Michalska and D. E. Webster, *Chem. Tech.*, 118 (1975).
37. J. C. Bailar, *Catal. Rev.*, **10**, 17 (1974).
38. K. G. Allum, R. D. Hancock, I. V. Howell, T. E. Lester, S. McKenzie, R. C. Pitkethly and P. J. Robinson, *J. Organometal. Chem.*, **107**, 393 (1976); K. G. Allum, R. D. Hancock, I. V. Howell, R. C. Pitkethly and P. J. Robinson, *J. Catal.*, **43**, 322 (1976).
39. A. A. Oswald and L. L. Murrell, *US Patent* 4,083,803 (1978).
40. I. Schwager and J. F. Knifton, *J. Catal.*, **45**, 256 (1976).
41. C. Y. Hsu and M. J. Orchin, *J. Am. Chem. Soc.*, **97**, 3553 (1975).
42. I. Wender, S. Metlin, S. Ergun, H. W. Sternberg and H. Greenfield, *J. Am. Chem. Soc.*, **78**, 5101 (1956).
43. B. Heil and L. Marko, *Chem. Ber.*, **102**, 2238 (1969).
44. A. J. M. Keulemans, A. Kwantes and T. van Bavel, *Recl. Trav. Chim. Pays-Bas*, **67**, 298 (1948).
45. A. J. de Jong and R. van Helden, *British Patent* 1,540,227 (1979) to Shell.
46. A. J. de Jong, *British Patent* 1,554,461 (1979) to Shell.
47. T. Shimizu, *German Patent* 2,538,364 (1976) to Kuraray.
48. C. U. Pittman and W. D. Honnick, *J. Org. Chem.*, **45**, 2132 (1980).
49. O. R. Hughes, *US Patents* 4,003,918 (1977) and 4,052,401 (1977) to Celanese.
50. A. M. Brownstein and H. L. List, *Hydrocarbon Process.*, **56** (9), 159 (1977).
51. C. C. Cumbo and K. K. Bhatia, *US Patent* 3,963,755 (1976) to du Pont.
52. P. Pino, C. Consiglio, C. Botteghi and C. Salomon, *Advan. Chem. Ser.*, **132**, 295 (1974).

53. P. Pino, F. Piacenti and M. Bianchi in *Organic Syntheses with Metal Carbonyls*, Vol. 2, (I. Wender and P. Pino, Eds.),Wiley, New York, 1977, p. 136.
54. M. Tanaka, Y. Ikeda and I. Ogata, *Chem. Letters*, 1158 (1975).
55. R. E. Merrill, *Chem. Tech.*, 118 (1981).
56. W. S. Knowles, M. J. Sabacky and B. D. Vineyard, *Advan. Chem. Ser.*, **132**, 274 (1974).
57. Y. Becker, A. Eisenstadt and J. K. Stille, *J. Org. Chem.*, **45**, 2145 (1980).
58. R. M. Laine, *J. Org. Chem.*, **45**, 3370 (1980) and references cited therein.
59. W. Reppe and H. Vetter, *Justus Liebig's Ann. Chem.*, **582**, 133 (1953).
60. J. Falbe, *J. Organometal. Chem.*, **94**, 213 (1975).

CHAPTER 5

OLEFIN CARBONYLATION AND RELATED REACTIONS

Carbon monoxide reacts with unsaturated compounds, such as olefins and acetylenes, in the presence of a variety of protic co-substrates and transition metal carbonyl catalysts to afford carboxylic acid derivatives (Reactions 1—6). These transformations are generally referred to as Reppe carbonylations after their discoverer [1].

$$HC\equiv CH \; + \; CO \; + \; ROH \longrightarrow H_2C=CHCO_2R \qquad (1)$$

$$HC\equiv CH \; + \; CO \; + \; R_2NH \longrightarrow H_2C=CHCONR_2 \qquad (2)$$

$$H_2C=CH_2 \; + \; CO \; + \; ROH \longrightarrow CH_3CH_2CO_2R \qquad (3)$$

$$H_2C=CH_2 \; + \; CO \; + \; R_2NH \longrightarrow CH_3CH_2CONR_2 \qquad (4)$$

$$H_2C=CH_2 \; + \; CO \; + \; RCO_2H \longrightarrow CH_3CH_2CO-O_2CR \qquad (5)$$

$$H_2C=CH_2 \; + \; CO \; + \; HCl \longrightarrow CH_3CH_2COCl \qquad (6)$$

$$R \; = \; H \text{ or alkyl}$$

A wide variety of protic co-substrates can be used, e.g. water, alcohols, amines and carboxylic acids. When the co-substrate is water the reaction is referred to as *hydroxycarbonylation* or hydrocarboxylation. The former is preferred, as is the term alkoxycarbonylation for the corresponding reaction with an alcohol as co-substrate. The reactions are carried out using catalytic amounts of Group VIII metal carbonyls under CO pressure or with stoichiometric amounts of metal carbonyls at atmospheric pressure. Addition of acids, such as hydrogen halides, accelerates the reaction (see below).

Initial work in this area was mainly concerned with the various carbonylation reactions of acetylene substrates, for which $Ni(CO)_4$ is the catalyst of choice [1—3]. For example, when acetylene is allowed to react with a mixture of carbon monoxide and water, in the presence of catalytic amounts of $Ni(CO)_4$ at 150°C and 30 bar, acrylic acid is formed in > 90% yield [1].

$$HC\equiv CH + CO + H_2O \xrightarrow{[Ni(CO)_4]} CH_2=CHCO_2H \qquad (7)$$

When mono-substituted acetylenes are used, carbonylation occurs predominantly at the substituted carbon to give a branched carboxylic acid, e.g.,

$$CH_3C\equiv CH + CO + H_2O \xrightarrow{[Ni(CO)_4]} CH_2=C(CH_3)CO_2H \qquad (8)$$

5.1 Olefin Carbonylation

The analogous hydroxycarbonylation of olefins affords a mixture of linear and branched carboxylic acids (Reaction 9). Whereas Ni is the preferred metal for acetylene carbonylation, catalysts based on Co, Pd, Rh, Fe and Ru have frequently been applied to olefin hydroxycarbonylations and related reactions [1–7]. The catalysts of choice for these reactions have generally been $Co_2(CO)_8$ [4, 5] or $(Ph_3P)_2PdCl_2$ [8–10]. These catalysts, in addition to being less toxic than $Ni(CO)_4$, are more active and afford higher selectivities to linear acids than nickel-based catalysts.

$$RCH=CH_2 + CO + H_2O \longrightarrow \begin{cases} RCH_2CH_2CO_2H \\ RCH(CH_3)CO_2H \end{cases} \qquad (9)$$

Compared to acetylene carbonylations more severe conditions of temperature and pressure are generally required for olefin carbonylations. It has been pointed out, however, that many early studies, in which high pressures were employed, were carried out at a time when the concepts of co-ordinative unsaturation and oxidative addition were not known [6]. Thus, the first step in many of these reactions is ligand dissociation to give reactive, co-ordinatively unsaturated species (see below). At high carbon monoxide pressures this step can be rate-limiting resulting in a decrease in rate with increasing pressure.

The hydroxycarbonylation of olefins can also be achieved using strong acid catalysts, such as H_2SO_4(Koch reaction). This results in the formation of branched acids via carbenium ion intermediates. Its synthetic utility is, in general, limited to the production of highly branched carboxylic acids, e.g. pivalic acid from isobutene [11].

$$(CH_3)_2C=CH_2 + CO + H_2O \xrightarrow{H^+} (CH_3)_3CCO_2H \qquad (10)$$

COMMERCIAL APPLICATION

Reppe carbonylations have, in general, had much less industrial impact than the related hydroformylation reaction. Thus, the estimated total capacities of hydroformylation, Reppe carbonylation and Koch carbonylation processes are 5.2 million, 600 000 and 60 000 tons per annum, respectively [3].

The first example of commercial utilization of Reppe carbonylations was the production of acrylic acid from acetylene (Reaction 7). This process, still operated by BASF, employs a nickel catalyst, at 40–55 bar and 180–200°C in tetrahydrofuran as solvent and gives acrylic acid selectivities of *ca.* 90%. The use of acetylene as a feedstock for the manufacture of acrylic acid and its esters is, however, gradually being displaced by propylene oxidation processes, mainly because of the high cost of acetylene.

Industrial application of olefin hydroxycarbonylation has been limited because good, alternative processes exist for the production of most large-volume carboxylic acids. However, some propionic acid is manufactured by the hydroxycarbonylation of ethylene (BASF process). The reaction employs a nickel propionate catalyst at 200–240 bar and 270–320°C in propionic acid as solvent. Propionic acid is formed in 95% yield [3].

$$CH_2=CH_2 + CO + H_2O \longrightarrow CH_3CH_2CO_2H \qquad (11)$$

Cobalt-catalysed methoxycarbonylations of long-chain olefins, such as 1-octene and internal dodecenes, have commercial utility as a source of fatty acid esters for use in synthetic lubricants or, after hydrogenation, detergent range alcohols [12].

$$RCH=CH_2 + CO + MeOH \xrightarrow[\text{pyridine}]{[Co_2(CO)_8]} RCH_2CH_2CO_2Me \qquad (12)$$

MECHANISM

The mechanism of the cobalt-catalysed hydroxycarbonylation of olefins is illustrated in Scheme 5.1. This mechanism was proposed by Heck and Breslow [13] and has many features in common with the related cobalt-catalysed hydro-formylation (see Chapter 4). The active catalyst is $HCo(CO)_4$, in this case formed by attack of water on co-ordinated CO in $Co_2(CO)_8$ in a water gas shift type reaction. The olefin then inserts into the Co–H bond to give an alkylcobalt species which subsequently undergoes CO insertion to afford an acylcobalt intermediate. Up till this point the mechanism is identical to that of the related hydroformylation reaction. The difference resides in the fate of the acylcobalt intermediate. In the Reppe reaction no hydrogen is available and the acylcobalt compound undergoes nucleophilic attack at the acyl group by water (or by an alcohol in alkoxycarbonylations). This results in cleavage to an acid (or ester) with concomitant regeneration of the $HCo(CO)_4$ catalyst.

The promoting action of hydrogen halides in many of these reactions is due to the facile formation of an active carbonylhydride species by oxidative addition of HX to a co-ordinatively unsaturated metal carbonyl. This is then followed by the usual sequence of olefin insertion, carbon monoxide insertion, etc., as

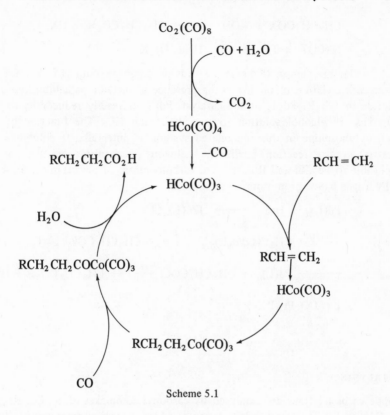

Scheme 5.1

illustrated in Scheme 5.2 for the nickel-catalysed alkoxycarbonylation of ethylene [14].

$$Ni(CO)_4 + HX \xrightarrow{-2\ CO} HNi(CO)_2X \qquad (13)$$

$$HNi(CO)_2X + CH_2{=}CH_2 \longrightarrow CH_3CH_2Ni(CO)_2X \qquad (14)$$

$$CH_3CH_2Ni(CO)_2X + CO \longrightarrow CH_3CH_2CONi(CO)_2X \qquad (15)$$

$$CH_3CH_2CONi(CO)_2X + ROH \longrightarrow CH_3CH_2CO_2R + HNi(CO)_2X \quad (16)$$

Scheme 5.2

The product-forming step (Reaction 16) as illustrated involves attack of the the alcohol on the acylnickel intermediate. A widely cited alternative involves a two-step reductive elimination of the acid chloride followed by alcoholysis:

$$CH_3CH_2CONi(CO)_2X \longrightarrow CH_3CH_2COX + Ni(CO)_2 \qquad (17)$$

$$CH_3CH_2COX + ROH \longrightarrow CH_3CH_2CO_2R + HX \qquad (18)$$

$$Ni(CO)_2 + HX \longrightarrow HNi(CO)_2X \qquad (19)$$

A similar mechanism (Scheme 5.3) was proposed by Tsuji [15–17] for the alkoxycarbonylation of olefins in the presence of metallic palladium/hydrogen chloride or $(Ph_3P)_2PdCl_2$ as the catalyst. $PdCl_2$ is readily reduced by carbon monoxide in alcohol solution to give Pd(0) and HCl. The function of the triphenylphosphine in this and related systems is, inter alia, to solubilize the catalyst in the reaction medium. Analogous mechanisms are presumably applicable to Fe, Ru and Rh-catalysed carbonylations of olefins in the presence of hydrogen halide promotors.

$$Pd(L)_n + HCl \longrightarrow HPd(L)_nCl$$

$$\xrightarrow{C_2H_4} CH_3CH_2Pd(L)_nCl \xrightarrow{CO} CH_3CH_2COPd(L)_nCl$$

$$\longrightarrow Pd(L)_n + CH_3CH_2COCl \xrightarrow{ROH} CH_3CH_2CO_2R + HCl$$

$$L = CO, Ph_3P$$

Scheme 5.3

CATALYSTS

Reppe carbonylations are catalysed by carbonyl complexes of Ni, Co, Rh, Pd, Pt, Ru and Fe. The metal carbonyls can be added as such or generated *in situ* by reacting either the finely divided metal or by reducing a metal salt with carbon monoxide. Above 100°C the metal carbonyls generally dissociate into the metal and CO unless a high pressure of CO is employed. However, with an appropriate choice of metal and ligand, catalysts can be obtained which are capable of effecting carbonylation under relatively mild conditions [6].

Nickel-catalysed carbonylations of linear alpha olefins afford the branched acid as the major product (60–70%), e.g.

$$CH_3CH{=}CH_2 + CO + H_2O \xrightarrow{Ni(CO)_4} (CH_3)_2CHCO_2H \qquad (20)$$

This contrasts with $Co_2(CO)_8$-catalysed carbonylations which give 55–60% linear acids, the total yield of carboxylic acids being 80–90% [3]. The selectivity to linear acids is increased to $> 80\%$ when pyridine is added as a co-catalyst. For example, Schaefer and co-workers [12] found that the methoxycarbonylation of 1-octene at 160 bar and 160°C, in the presence of $Co_2(CO)_8$-pyridine, afforded methyl nonanoate in 98% selectivity with a linearity of *ca.* 85%.

$$CH_3(CH_2)_5 CH{=}CH_2 + CO + MeOH \xrightarrow[\text{pyridine}]{[Co_2(CO)_8]} CH_3(CH_2)_7 CO_2 Me \qquad (21)$$

The active catalyst in this system is presumably the pyridinium salt $C_5H_5NH^+Co(CO)_4^-$. Since cobalt carbonyls readily catalyse the isomerization of olefins the carbonylation of internal olefins generally affords the linear acid (c.f. cobalt-catalysed hydroformylation). Thus, methoxycarbonylation of a mixture of internal dodecenes using the $Co_2(CO)_8$-pyridine system affords the linear C_{13} carboxylic acid ester as the major product [12]. Since linear acids (esters) are generally the desired products the high degree of linearity afforded by cobalt catalysts is an important advantage over, for example, nickel catalysts.

Palladium salts have been extensively studied as carbonylation catalysts [7–10, 15–17]. When $PdCl_2$ [15–17] or $(Ph_3P)_2PdCl_2$ [8–10] are used, linear alpha olefins yield predominantly the branched acid (ester). In contrast, the combination $(R_3P)_2 PdCl_2-SnCl_2$ affords linear esters in 85–90% selectivity in the methoxycarbonylation of linear alpha olefins at 70°C and 136 bar [9]. Branched alpha olefins also afford linear esters (Reaction 22) whilst internal olefins yield branched esters as the major products. The latter result contrasts with that observed with cobalt catalysts (see above).

$$CH_3(CH_2)_2 C(CH_3){=}CH_2 + CO + MeOH$$

$$\xrightarrow{(Ph_3P)_2PdCl_2-SnCl_2} CH_3(CH_2)_2 CH(CH_3)CH_2 CO_2 Me \quad 99\% \text{ selectivity} \quad (22)$$

The active catalyst in this system is probably $HPd(SnCl_3)(R_3P)_2$ and the selectivity for linear esters was attributed to the steric effect of the phosphine and $SnCl_3^-$ ligands [9]. Similarly, platinum-tin catalysts are also highly selective for the conversion of linear alpha olefins to linear esters [18]. They are, however, significantly less active than the analogous palladium catalysts and are completely ineffective with branched and internal olefins.

Bearing in mind the superior activity of rhodium catalysts in olefin hydroformylation and alcohol carbonylation (Chapter 7) it is rather surprising that rhodium has received scant attention as an olefin carbonylation catalyst. The rhodium-catalysed hydroxycarbonylation of ethylene, in the presence of HBr or HI promotors, at 25–50 bar and 175°C, gives propionic acid in good yield [19]. Higher alpha olefins, such as 1-hexene, afford carboxylic acids with 40–65% linearity [19].

OLEFIN SUBSTRATES

Olefin carbonylations generally exhibit the same structure-reactivity pattern as the closely related hydroformylation reaction. Thus, linear alpha olefins react the most readily and the rate decreases with increasing substitution at the double bond. Whether branched or linear acids (esters) are the major products depends very much on the catalyst system employed (see above).

An interesting synthetic application of olefin carbonylation is the alkoxy-carbonylation of α, β-unsaturated nitriles to give the corresponding α-cyano-carboxylic acid esters [20]. For example, $Co_2(CO)_8$-catalysed ethoxycarbonylation of acrylonitrile in the presence of α-picoline as promotor at 125°C and 140 bar gave α-cyanopropionic acid ethyl ester in 95% yield.

$$RCH=CHCN + CO + EtOH \longrightarrow RCH_2CH(CN)CO_2Et \qquad (23)$$

R	Yield (%)
H	95
CH_3	95
Ph	60
CN	75

Recently there has been a growing interest in the carbonylation of conjugated dienes, and butadiene in particular. The product observed depends very much on the catalyst system used. For example, cobalt-catalysed methoxycarbonylation of butadiene at 135°C and 900 bar affords dimethyl adipate [21].

$$CH_2=CH-CH=CH_2 + CO + MeOH \xrightarrow[C_5H_5N]{[Co_2(CO)_8]} MeO_2C(CH_2)_4CO_2Me \quad (24)$$

This reaction has potential commercial utility as a route to adipic acid. The direct hydroxycarbonylation of butadiene to adipic acid using an iodide-promoted rhodium catalyst at 75 bar and 200°C has also been reported [22].

$$CH_2=CH-CH=CH_2 + CO + H_2O \xrightarrow[HI \text{ or } HBr]{[RhCl_3]} HO_2C(CH_2)_4CO_2H \qquad (25)$$

The outcome of the palladium-catalysed alkoxycarbonylation of butadiene depends on the presence or absence of halide ions. $PdCl_2$-catalysed carbonylation [23] affords the 3-pentenoic acid ester (Reaction 26). By contrast, halide-free catalysts, such as $(Ph_3P)_2Pd(OAc)_2$ [24, 25], give the 3,8-nonadienoic acid ester via a combined dimerization-carbonylation (Reaction 27).

$$CH_2=CH-CH=CH_2 + CO + ROH \begin{cases} \xrightarrow{[PdCl_2]} CH_3CH=CHCH_2CO_2R & (26) \\ \xrightarrow{[(Ph_3P)_2Pd(OAc)_2]} CH_2=CH(CH_2)_3CH=CHCH_2CO_2R & (27) \end{cases}$$

OXIDATIVE CARBONYLATION

When palladium-catalysed carbonylations are carried out in the presence of an oxidant this leads to *oxidative carbonylation* [26–28]. For example, the $PdCl_2$–$CuCl_2$ catalysed carbonylation of ethylene in the presence of oxygen leads to the selective formation of acrylic acid (Reaction 28). This reaction is of interest as an ethylene-based route to acrylic acid.

$$CH_2{=}CH_2 + CO + \tfrac{1}{2}O_2 \xrightarrow{[Pd^{II}, Cu^{II}]} CH_2{=}CHCO_2H \qquad (28)$$

The mechanism proposed [26] for this reaction involves as a key feature the attack of water on co-ordinated CO to afford a co-ordinated carboxyl group followed by olefin insertion:

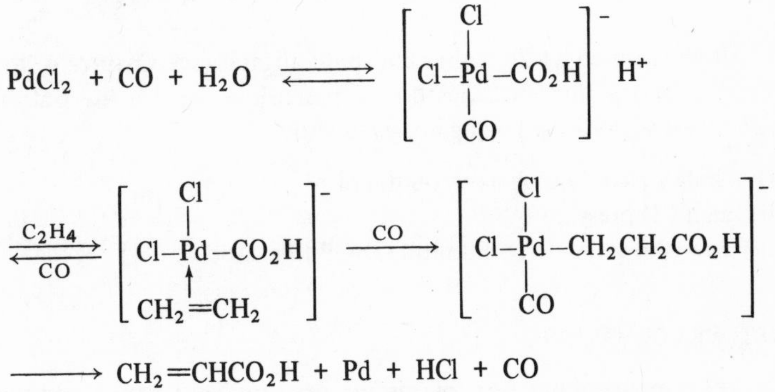

$$PdCl_2 + CO + H_2O \rightleftharpoons \left[\begin{array}{c} Cl \\ | \\ Cl-Pd-CO_2H \\ | \\ CO \end{array} \right]^{-} H^{+}$$

$$\xrightarrow[\text{CO}]{C_2H_4} \left[\begin{array}{c} Cl \\ | \\ Cl-Pd-CO_2H \\ \uparrow \\ CH_2{=}CH_2 \end{array} \right]^{-} \xrightarrow{\text{CO}} \left[\begin{array}{c} Cl \\ | \\ Cl-Pd-CH_2CH_2CO_2H \\ | \\ CO \end{array} \right]^{-}$$

$$\longrightarrow CH_2{=}CHCO_2H + Pd + HCl + CO$$

Scheme 5.4

The function of the copper co-catalyst is to mediate the re-oxidation of the palladium catalyst via:

$$Pd^0 + 2CuCl_2 \longrightarrow PdCl_2 + 2CuCl \qquad (29)$$

$$2CuCl + \tfrac{1}{2}O_2 + 2\ HCl \longrightarrow 2CuCl_2 + H_2O \qquad (30)$$

The oxidative alkoxycarbonylation of ethylene to succinate esters (Reaction 31) has also been reported [29].

$$CH_2{=}CH_2 + 2\ CO + 2\ ROH + \tfrac{1}{2}O_2 \xrightarrow{[Pd^{II}, Cu^{II}]}$$
$$RO_2CCH_2CH_2CO_2R + H_2O \qquad (31)$$

Similarly, the oxidative alkoxycarbonylation of butadiene with a $PdCl_2$–$CuCl_2$ catalyst, in the presence of a stoichiometric amount of a dehydrating agent (to remove the water formed), has recently been reported [30].

$$CH_2{=}CH{-}CH{=}CH_2 + 2\ ROH + 2\ CO + \tfrac{1}{2}O_2$$

$$\xrightarrow{[Pd^{II}, Cu^{II}]} RO_2CCH_2CH{=}CHCH_2CO_2R + H_2O \qquad (32)$$

ASYMMETRIC CARBONYLATION

The carbonylation of prochiral olefins in the presence of catalysts containing appropriate chiral ligands leads to asymmetric induction. One system that has been studied in detail is the palladium-catalysed alkoxycarbonylation of α-methylstyrene [31]. The system is a favourable one for study as the linear 3-phenylbutanoate ester is formed in > 95% yield.

$$PhC(CH_3){=}CH_2 + CO + ROH \xrightarrow{[Pd(DIOP)]} Ph\overset{*}{C}H(CH_3)CH_2CO_2R \quad (33)$$

$$59\% \text{ e.e.}$$

DIOP was used as the chiral phosphine ligand (see Chapter 4, for structure). Changes in reaction variables had a dramatic effect on the optical yield, the highest value (59% e.e.) being observed with:

(i)   a bulky alcohol such as tert-butanol
(ii) high CO pressure
(iii) low phosphine to metal ratio (DIOP : Pd ratio of 0.4)

KETONE FORMATION

The hydroxycarbonylation of olefins can, under certain conditions, lead to ketone formation. For example, the conversion of ethylene to diethyl ketone (Reaction 34) occurs with both rhodium [32] and ruthenium [33] catalysts. The yield of diethylketone was 81% (98% selectivity on converted ethylene) using a $RuCl_3$ catalyst at 20 bar and $80°C$ [33].

$$2\ C_2H_4 + 2\ CO + H_2O \longrightarrow (C_2H_5)_2CO + CO_2 \quad (34)$$

5.2 Carbonylation of Organic Halides

In addition to olefin carbonylations a variety of related carbon monoxide insertion reactions are catalysed by Group VIII metal complexes. Most of these reactions were reported by Reppe in his seminal study of catalytic carbonylation processes [1]. The full scope of their synthetic utility is, however, just beginning to be appreciated. The carbonylations of alcohols, ethers and esters are dealt with in Chapter 7 and nitrogen-containing substrates are treated in Chapter 8. Another important class of reactions involves the insertion of CO into a carbon-halogen bond.

The carbonylation of organic halides is catalysed by low-valent complexes of Ni, Co, Fe, Rh and Pd. The reaction generally occurs under much milder conditions than the olefin carbonylations described in the preceding section.

It involves oxidative addition, carbon monoxide insertion and reductive elimination steps according to the general scheme:

$$M(L)_n + RX \longrightarrow R-M(L)_n X \tag{35}$$

$$R-M(L)_n X + CO \longrightarrow RCOM(L)_n X \tag{36}$$

$$RCOM(L)_n X \longrightarrow RCOX + M(L)_n \tag{37}$$

$$L = CO, Ph_3P, \text{ etc.} \qquad X = Cl, Br, I, RSO_3, \text{ etc.}$$

Scheme 5.5

The acylmetal intermediate undergoes solvolysis *in situ* by a variety of protic co-substrates, e.g. water, amines, thiols to afford carboxylic acid derivatives:

$$RCOMX \begin{cases} \xrightarrow{H_2O} RCO_2H + HX + M & (38) \\ \xrightarrow{R'OH} RCO_2R' + HX + M & (39) \\ \xrightarrow{R'_2NH} RCONR'_2 + HX + M & (40) \end{cases}$$

The reaction is most favourably carried out using a metal carbonyl, or a precursor thereof, in the presence of a stoichiometric amount of a base such as a tertiary amine or a metal alkoxide, leading to the overall reaction:

$$RX + CO + R'O^- \xrightarrow{[Co_2(CO)_8]} RCO_2R' + X^- \tag{41}$$

For example, the $Co_2(CO)_8$-catalysed alkoxycarbonylation of alkyl and benzyl halides has been extensively studied [13, 34]. The reaction was first described by Heck and Breslow [13] who showed that it involves nucleophilic displacement of halide by the tetracarbonylcobaltate anion, $Co(CO)_4{}^-$, and subsequent CO insertion into the alkylcobalt intermediate:

$$RX + Co(CO)_4{}^- \longrightarrow RCo(CO)_4 + X^- \tag{42}$$

$$RCo(CO)_4 + CO \longrightarrow RCOCo(CO)_4 \tag{43}$$

$$RCOCo(CO)_4 + R'O^- \longrightarrow RCO_2R' + Co(CO)_4{}^- \tag{44}$$

The observed structure-reactivity pattern − R = benzyl, allyl > primary alkyl > secondary alkyl > tertiary alkyl − is consistent with rate-limiting S_N2 displacement of halide by $Co(CO)_4{}^-$ (Reaction 42). Aryl and vinyl halides are, as expected for an S_N2 mechanism, unreactive. The carbonylation of these halides can be achieved, however, with Ni and Pd catalysts (see below). In these cases activation via π-complex formation is probably involved.

BENZYLIC HALIDES

Recently, El-Chahawi and co-workers [34–37] have developed the $Co_2(CO)_8$-catalysed carbonylation of benzylic chlorides into a synthetically useful method for the preparation of phenylacetic esters (Reaction 45).

$$ArCH_2Cl + CO + NaOMe \xrightarrow{[Co_2(CO)_8]} ArCH_2CO_2Me + NaCl \qquad (45)$$

The reaction proceeds readily at 55°C and 1–2 bar CO pressure to afford phenylacetic esters in excellent yields (see Table 5.I). The catalyst can be added as $Co_2(CO)_8$ or generated *in situ* by reduction of a cobalt(II) salt with a mixture of manganese powder and sodium dithionite [35].

TABLE 5.I

$Co_2(CO)_8$-catalysed methoxycarbonylation of benzyl chlorides[a]

$$ArCH_2Cl + CO + NaOMe \longrightarrow ArCH_2CO_2Me + NaCl$$

Ar	$ArCH_2CO_2Me$ yield (%)	Reference
phenyl	95	35
p-chlorophenyl	90	35
p-methylphenyl	87	35
p-methoxyphenyl	32	35
p-isopropylphenyl	64	35
p-cyanophenyl	30	34
p-nitrophenyl	10	34
o-methylphenyl	75	34
m-methylphenyl	79	34
o-carboxyphenyl	74	37
m-carboxyphenyl	78	37
p-carboxyphenyl	89	37

[a] Typical conditions: 3g $Co_2(CO)_8$, 0.5 mol $ArCH_2Cl$, 0.5 mol 25% NaOMe in MeOH treated with CO at 55°C and 1–2 bar for *ca.* 5 hr.

Reaction 45 is also catalysed by $Fe(CO)_5$ [35] and $RhCl_3$ [38] and an analogous alkoxycarbonylation of benzyl halides is observed with $(Ph_3P)_2PdCl_2$ in the presence of stoichiometric amounts of a tertiary amine [39]. However, cost-efficiency considerations make $Co_2(CO)_8$ the catalyst of choice (see Table 5.II for a comparison of various catalysts).

Reaction 45 constitutes a commercially attractive alternative to the conventional route for converting benzyl chlorides to phenylacetic esters, which involves reaction with cyanide ion followed by alcoholysis. A commercial

Table 5.II

Catalysts for the methoxycarbonylation of benzyl chloride at 55°C

Catalyst	Catalyst/PhCH$_2$Cl molar ratio	Pressure (bar)	Time (h)	Yield (%)	Reference
Fe(CO)$_5$	1/9.5	1.7	6	70	35
Co$_2$(CO)$_8$	1/86	1.7	6	95	35
Co$_2$(CO)$_8$	1/272	6	2	95	36
RhCl$_3$	1/395	1.7	6	93	38

process for the manufacture of methyl phenylacetate using the Co$_2$(CO)$_8$ system has been described [34].

Similarly, substituted benzyl chlorides afford the corresponding α-substituted phenylacetic esters [40] and *bis*-chloromethylbenzenes are converted to dicarboxylic acid esters [41].

$$PhCH(CH_3)Cl + CO + NaOMe \xrightarrow[\text{1.8 bar, 55°C}]{[Co_2(CO)_8]} PhCH(CH_3)CO_2Me + NaCl \quad (46)$$

51% yield

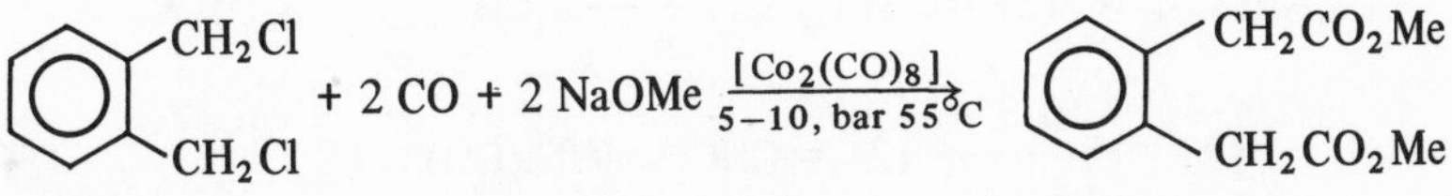

(47)

50% yield

The reaction is not limited to benzylic chlorides. Thus, chloromethylthiophene is converted to α-thienylacetic esters in good yield [42, 43].

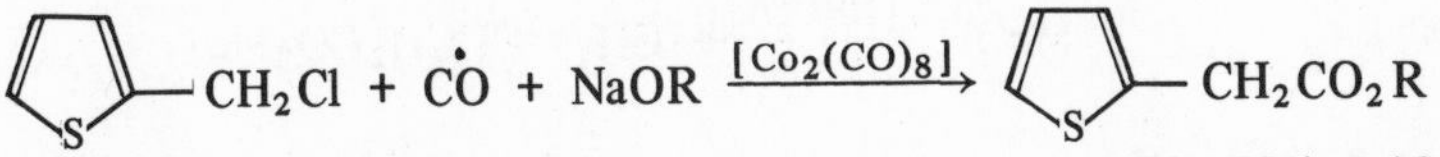

(48)

70–75% yield

Under more forcing conditions (80°C, 150 bar) benzal halides are converted to phenylmalonic esters using triethylamine as the base [44].

$$PhCHCl_2 + 2 CO + 2 EtOH \xrightarrow[\text{Et}_3\text{N}]{[Co_2(CO)_8]} PhCH(CO_2Et)_2 \quad (49)$$

62% yield

When the Co$_2$(CO)$_8$-catalysed carbonylation of benzyl chlorides is carried out in *tert*-butanol at low temperatures (35°C) and elevated pressures (20 bar CO), using LiOH as the base, a double insertion of CO is observed and the phenylpyruvic acid ester is formed in high yield [45].

$$PhCH_2Cl \xrightarrow[(2)]{(1)} PhCH_2COCO_2H \qquad\qquad (50)$$

$$90\% \text{ yield}$$

(1) $CO/t\text{-BuOH}/Co_2(CO)_8/LiOH$
(2) HCl

The $Co_2(CO)_8$-catalysed carbonylation of benzyl chlorides in aqueous methanol solutions of NaOH affords the corresponding phenylacetic acid (sodium salt) in high yield [46]. This reaction can be advantageously carried out in a two-phase, aqueous NaOH/hydrocarbon solvent system with the aid of a phase transfer catalyst (see below).

ALLYLIC HALIDES

The preferred catalyst for the alkoxycarbonylation of allylic halides is $Ni(CO)_4$ [6, 34, 47]. These reactions involve π-allylnickel complexes as reactive intermediates [14]:

$$Ni(CO)_n + CH_2{=}CHCH_2X \longrightarrow \underset{\text{CH}_2}{\overset{\text{CH}_2}{CH}} \hspace{-1em} {:}Ni(CO)_nX$$

$$\longrightarrow CH_2{=}CHCH_2CONi(CO)_{n-1}X \qquad\qquad (51)$$

In the presence of thiourea as promotor the methoxycarbonylation of allyl chloride proceeds readily at atmospheric pressure and ambient temperature to give methyl 3-butenoate in 89% yield [48].

$$CH_2{=}CHCH_2Cl + CO + MeOH \xrightarrow[\text{thiourea}]{[Ni(CO)_4]} CH_2{=}CHCH_2CO_2Me \qquad (52)$$

In contrast, when the reaction is carried out with sodium methoxide in methanol at pH 8–10, in the absence of a promotor at $60°C$ and 1–2 bar, methyl crotonate is formed in 90% yield [49]. The same reaction is observed with $Co_2(CO)_8$ as the catalyst albeit in lower yield (73% selectivity at 74% conversion).

$$CH_2{=}CHCH_2Cl + CO + NaOMe \xrightarrow{[Ni(CO)_4]} CH_3CH{=}CHCO_2Me + NaCl \qquad (53)$$

ALKYL HALIDES

Alkyl halides are less reactive than benzylic and allylic halides and have received less attention. It is worth noting, however, that the anti-Markownikov (free radical) addition of HBr to an alpha olefin followed by carbonylation of the resulting primary alkyl bromide (Reaction 54) offers a potentially useful alternative to conventional olefin carbonylation. As noted earlier, the disadvantage of the latter is that it produces isomeric mixtures. Alkyl halide carbonylations, on the other hand, if carried out at relatively low temperatures ($< 100°C$), afford only the expected isomer [13].

$$RCH{=}CH_2 + HBr \longrightarrow RCH_2CH_2Br \xrightarrow[[Co_2(CO)_8]]{CO/ROH} RCH_2CH_2CO_2R \quad (54)$$

An interesting commercial application of alkyl halide carbonylation is the Dynamit Nobel process for the manufacture of malonic esters by $Co_2(CO)_8$-catalysed alkoxycarbonylation of chloroacetic acid esters [34, 50]. The reaction proceeds under mild conditions (55°C and 7 bar) and, by controlling the pH in the range 7–8.5, excellent yields of malonic esters are obtained.

$$ClCH_2CO_2R + CO + NaOR \xrightarrow{[Co_2(CO)_8]} CH_2(CO_2R)_2 + NaCl \quad (55)$$

R	$CH_2(CO_2R)_2$ yield (%)
Me	91
Et	95
i-Pr	92

The higher reactivity of $ClCH_2CO_2R$ and $ClCH_2CN$ [51] compared to simple alkyl chlorides is consistent with their increased reactivity towards S_N2 displacement resulting from the electron-withdrawing effect of the CO_2R (and CN) group.

Alkyl halides have also been converted to carboxylic acids and esters using $Na_2Fe(CO)_4$ in stoichiometric quantities [52]. A more recent variant involves the use of $FeCl_3$ in combination with t-AmONa/NaH/CO for the $in\text{-}situ$ generation of $Na_2Fe(CO)_4$. This system was used for the carbonylation of primary, secondary and tertiary alkyl halides at 20–70°C and atmospheric pressure [53]. The reaction could also be carried out using catalytic amounts of $FeCl_3$. It is noteworthy that even tertiary alkyl halides can be converted to give, after acidic work-up, carboxylic acids in good yields. Since S_N2 substitution of tertiary alkyl halides is unlikely carbonylation of such substrates probably involves an electron transfer mechanism (Scheme 5.6).

$$RX + Fe(CO)_4{}^{2-} \longrightarrow R{\cdot} + Fe(CO)_4{}^- + X^-$$

$$\longrightarrow RFe(CO)_4{}^- \xrightarrow{CO} RCOFe(CO)_4{}^-, \text{etc.}$$

Scheme 5.6

ARYL HALIDES

The system $RONa/NaH/Co(OAc)_2/CO$ in tetrahydrofuran was similarly used for the *in situ* generation, at ambient temperature and atmospheric pressure, of a cobalt carbonylation catalyst of unprecedented reactivity [54, 55]. For example, this reagent catalysed the carbonylation of aryl halides, at atmospheric pressure and 60°C, to afford mixtures of the benzoic acids and esters in good yield [55]. This reaction presumably involves an electron transfer ($S_{RN}1$) mechanism and aryl radicals as transient intermediates (Scheme 5.7). The observed formation of 2-aryltetrahydrofurans as by-products is consistent with such a mechanism.

$$ArX \xrightarrow[\text{or } RO^-]{Co(CO)_n^-} [ArX]^{\cdot-} \longrightarrow Ar\cdot + X^-$$

$$\xrightarrow{Co(CO)_n^-} ArCo(CO)_n^- \xrightarrow{CO} ArCOCo(CO)_n^-, \text{etc.}$$

Scheme 5.7

The carbonylation of vinyl and aryl halides is not catalysed by cobalt carbonyl under normal conditions, consistent with the inertness of such halides to S_N2 displacement by $Co(CO)_4^-$ (Reaction 42). The carbonylation of these halides is, however, readily catalysed by low-valent complexes of nickel and palladium. The latter undergo facile oxidative addition reactions with vinyl and aryl halides [56], e.g.,

$$ArX + (Ph_3P)_4Pd \longrightarrow ArPd(Ph_3P)_2X + 2\ Ph_3P \qquad (56)$$

Such reactions may involve, as the initial step, a single electron transfer from palladium(0) to the aromatic halide with formation of the aromatic radical anion as a transient intermediate (c.f. Scheme 5.7). In the presence of carbon monoxide migratory insertion into the Ar—Pd bond occurs and palladium catalysts have been widely used for the alkoxycarbonylation of aryl halides in the presence of a tertiary amine as base [39, 57, 58].

$$ArX + ROH + R'_3N + CO \xrightarrow[\text{or } L_2PdCl_2]{L_4Pd} ArCO_2R + R'_3NHX \qquad (57)$$

ALDEHYDE FORMATION

When the carbonylation of organic halides is carried out using syn gas instead of carbon monoxide this can result in the selective formation of aldehydes. For example, the reaction of benzyl chloride with syn gas ($CO/H_2 = 1 : 1$) in the presence of $Co_2(CO)_8$ and dimethylformamide, in acetone solvent at 110°C and

200 bar gives phenylacetaldehyde in 52% yield [59]. Substituted benzyl halides were similarly converted to the corresponding phenylacetaldehydes using the same system.

$$PhCH_2Cl + CO + H_2 \xrightarrow[DMF]{[Co_2(CO)_8]} PhCH_2CHO + HCl \qquad (58)$$

A plausible mechanism for Reaction 58 involves the hydrogenolysis of an acylcobalt intermediate as the product-forming step (Scheme 5.8). The dimethylformamide presumably functions as a base in this system.

$$ArCH_2Cl + Co(CO)_4^- \longrightarrow ArCH_2Co(CO)_4 + Cl^-$$

$$\xrightarrow{CO} ArCH_2COCo(CO)_4 \xrightarrow[\text{or } H_2]{HCo(CO)_4} ArCH_2CHO$$

Scheme 5.8

Similarly, the reaction of aryl halides with syn gas ($CO/H_2 = 1 : 1$) in the presence of a tertiary amine and catalytic amounts of $(Ph_3P)_2PdCl_2$, at 80 bar and 100°C, affords the corresponding benzaldehyde [60].

$$ArX + CO + H_2 + R_3N \xrightarrow{[L_2PdCl_2]} ArCHO + R_3NHX \qquad (59)$$

The active catalyst is presumed to be $Pd(CO)L_2$, formed in situ by reduction with CO, and the catalytic cycle can be represented as:

$$ArX + Pd(CO)L_2 \longrightarrow ArPd(CO)L_2X$$

$$\longrightarrow ArCOPdL_2X \xrightarrow{H_2} ArCHO + HPdL_2X$$

$$\xrightarrow[CO]{R_3N} R_3NHX + Pd(CO)L_2, \text{ etc}$$

Scheme 5.9

PHASE TRANSFER CATALYSIS

A recent development that has enormous potential in industrial organic synthesis is the application of phase transfer catalysis to metal-catalysed carbonylations. The technique of phase transfer catalysis [61, 62] allows reactions of inorganic salts with water-insoluble organic substrates to be carried out in two-phase aqueous-organic media. The phase transfer catalyst — a tetraalkylammonium or phosphonium salt or a crown ether — effectively shuttles anions between the aqueous and organic phases. For example, in the reaction of cyanide ion with an alkyl chloride (Scheme 5.10) the cyanide ion is solubilised in the organic phase as the tetraalkylammonium cyanide, $R_4N^+CN^-$. Reaction between the latter and the alkyl chloride, in the organic phase, affords the alkyl cyanide and $R_4N^+Cl^-$ which is transferred back to the aqueous phase where it undergoes exchange with sodium cyanide to regenerate $R_4N^+CN^-$.

Aqueous phase $NaCN + R_4NCl \rightleftharpoons NaCl + R_4NCN$

Interface $- - - - - - - \uparrow - - - - - - - - - - - \downarrow - - -$

Organic phase $R'CN + R_4NCl \longleftarrow R'Cl + R_4NCN$

Scheme 5.10

Phase transfer catalysed reactions possess several advantages as a synthetic tool. They generally proceed rapidly under mild conditions without the use of expensive solvents and they are easy to work up. It is a technique of considerable industrial importance which has been extensively applied in organic chemistry [61, 62]. More recently, attention has been focussed on its utility in organometallic chemistry and in metal-catalysed processes [63, 64].

The technique has been applied particularly successfully to catalytic carbonylations. For example, the $Co_2(CO)_8/R_4NX$-catalysed hydroxycarbonylation of benzylic halides, in a two-phase water/hydrocarbon mixture, proceeds readily at atmospheric pressure and $20-50°C$ to give the corresponding phenylacetic acids in high yields [63–70].

$$ArCH_2Cl + CO + 2NaOH \xrightarrow[H_2O/ArH]{[Co_2(CO)_8/R_4NX]} ArCH_2CO_2Na + NaCl + H_2O \quad (60)$$

The method is also applicable to alkyl halides [64] and has broad synthetic utility. The active catalyst is the tetracarbonylcobaltate anion which is present in the organic phase as $R_4N^+Co(CO)_4^-$ where the following reactions take place:

$$R_4N^+Co(CO)_4^- + ArCH_2Cl \longrightarrow R_4NCl + ArCH_2Co(CO)_4 \quad (61)$$

$$ArCH_2Co(CO)_4 + CO \longrightarrow ArCH_2COCo(CO)_4 \quad (62)$$

$$ArCH_2COCo(CO)_4 + R_4NOH \longrightarrow R_4N^+ArCH_2CO_2^- + HCo(CO)_4 \quad (63)$$

The carboxylate anion is subsequent transferred to the aqueous phase where it exchanges with sodium hydroxide to regenerate R_4NOH. Reaction of the latter with $HCo(CO)_4$ completes the catalytic cycle.

$$R_4N^+ArCH_2CO_2^- + NaOH \rightleftharpoons ArCH_2CO_2Na + R_4NOH \quad (64)$$

$$R_4NOH + HCo(CO)_4 \longrightarrow R_4N^+Co(CO)_4^- + H_2O \quad (65)$$

The $R_4N^+Co(CO)_4^-$ resides predominantly in the organic phase and when the reaction is complete the catalyst solution is separated from the aqueous phase which contains the product as a sodium salt. Thus, simple phase separation allows for facile recycling of the catalyst, an important advantage in industrial applications.

Interestingly, tetraalkylammonium salts can also act as substrates in these carbonylation reactions [71]. In this case the key step probably involves S_N2 displacement of R_3N as the leaving group instead of halide.

$$ArCH_2\overset{+}{N}R_3 + Co(CO)_4^- \longrightarrow ArCH_2Co(CO)_4 + R_3N \quad (66)$$

Alternatively, a single electron transfer mechanism with benzyl radicals as obligatory intermediates cannot be ruled out on the basis of the experimental evidence.

$$ArCH_2\overset{+}{N}R_3 + Co(CO)_4^- \longrightarrow ArCH_2\cdot + Co(CO)_4 + R_3N$$
$$\longrightarrow ArCH_2Co(CO)_4, \text{etc} \quad (67)$$

The phase transfer catalysed, $Co_2(CO)_8$-mediated carbonylation (1 bar CO) of aryl and vinyl halides under photostimulation (350 nm) affords the corresponding carboxylic acids in high yield [72].

$$ArX + CO + 2NaOH \xrightarrow[\substack{PhH/H_2O \\ h\nu\,(350\,nm)}]{[Co_2(CO)_8/Bu_4NBr]} ArCO_2Na + NaX + H_2O \quad (68)$$

The palladium-catalysed carbonylations of benzyl and aryl halides can also be carried out conveniently under phase transfer conditions [64, 73, 74]. For example, the $(Ph_3P)_2PdCl_2/PhCH_2NEt_3Cl$-catalysed carbonylation of benzyl chloride, in p-xylene/aqueous NaOH mixtures at 95°C, affords phenylacetic acid in 80–90% yield [64, 73].

The hydroxycarbonylation of aryl halides can be carried out under similar conditions [64, 74]. Bromobenzene, for example, is converted to benzoic acid in 86% yield [74]. This technique is very effective for the selective mono-carbonylation of polyhalogenated aromatics. p-Dibromobenzene, for example, is converted to p-bromobenzoic acid in 95% yield [64].

$$Br - \langle\bigcirc\rangle - Br + CO + 2NaOH$$
$$\xrightarrow{[L_4Pd/R_4NX]} Br - \langle\bigcirc\rangle - CO_2Na + NaBr \quad (69)$$

1,3,5-Trichlorobenzene is similarly converted to 3,5-dichlorobenzoic acid in 97% selectivity [64].

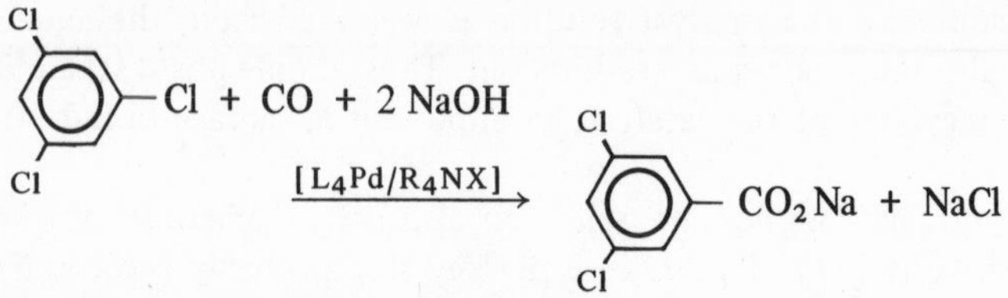

$$\text{Cl}\underset{\text{Cl}}{\overset{\text{Cl}}{\bigcirc}}\!\!-\text{Cl} + \text{CO} + 2\,\text{NaOH} \xrightarrow{[\text{L}_4\text{Pd}/\text{R}_4\text{NX}]} \underset{\text{Cl}}{\overset{\text{Cl}}{\bigcirc}}\!\!-\text{CO}_2\text{Na} + \text{NaCl} \qquad (70)$$

Such high selectivities are possible because carbonylation of the first C–X bond results in the transfer of the product (ArCO_2^-) to the aqueous phase, thus removing it from the sphere of reaction.

The application of phase transfer catalysis to catalytic carbonylations is in a preliminary stage of development. However, judging from the successes already booked this technique would appear to have a promising future. Its advantages can be summarized as:

— The reactions take place under mild conditions, due to the enhanced reactivity of the organometallic intermediates in non-polar media.
— Easy removal and recycle of the catalyst.
— Selective monocarbonylation of polyhalogenated substrates.

It is worth noting that the technique has not yet been applied to the olefin hydroxycarbonylations discussed earlier.

5.3 Aromatic Carbonylation

The metal carbonyl-catalysed carbonylation of aromatics, analogous to the corresponding reaction with olefins described earlier, is not known. The lack of reactivity of aromatic nuclei in this reaction is due to their inability to insert readily into metal hydride bonds to form the required arylmetal intermediate.

The carbonylation of aromatics does occur, however, in the presence of strong Lewis or Bronsted acid catalysts.

In the classic Gatterman–Koch aldehyde synthesis (Reaction 71) aromatic hydrocarbons react with carbon monoxide in the presence of stoichiometric amounts of aluminium chloride and hydrogen chloride [75]. The need for stoichiometric amounts of aluminium chloride is dictated by the fact that it forms a strong complex with the aldehyde product. This complex is decomposed at the end of the reaction.

$$\text{ArH} + \text{CO} \xrightarrow{\text{AlCl}_3/\text{HCl}} \text{ArCHO} \qquad (71)$$

Two modes of operation are known: low (atmospheric) and high pressure synthesis (up to 100–200 bar). Under atmospheric pressure conditions cuprous chloride is added in order to facilitate, through complexation, absorption of the CO into the reaction medium. At high pressures no carrier is needed. Excess aromatic substrate is generally employed as the solvent and the reaction temperature is usually $< 50°C$, since higher temperatures lead to side reactions.

Benzene affords benzaldehyde in yields up to 90%. Toluene and chlorobenzene are converted to p-tolualdehyde (85% yield) and p-chlorobenzaldehyde (70% yield), respectively. The scope of the reaction is, however, limited since higher alkylbenzenes generally undergo scrambling via alkylation–dealkylation steps. The Gatterman–Koch reaction is essentially a Friedel–Crafts acylation of aromatic hydrocarbon with formyl chloride, formed as a transitory intermediate from carbon monoxide and hydrogen chloride.

$$CO + HCl \rightleftharpoons HCOCl \tag{72}$$

$$HCOCl + ArH \xrightarrow{AlCl_3} ArCHO + HCl \tag{73}$$

The carbonylation of aromatics also proceeds smoothly in the presence of HF/BF_3 (modified Gatterman–Koch synthesis). Although the Gatterman–Koch reaction has been known since the last century and its modified version since 1949 [76], it is only recently that this technology has been applied commercially. The most serious obstacle to commercialization was the difficulty of recycling the HF/BF_3 (or $AlCl_3$). This problem has been overcome by Mitsubishi Gas who have developed a commercial process for the manufacture of p-tolualdehyde by HF/BF_3-catalysed carbonylation of toluene [77]. This is the first step in a two-step process for the conversion of toluene to p-terephthalic acid.

$$\tag{74}$$

$$\tag{75}$$

This process constitutes a commercially viable alternative to existing processes based on p-xylene oxidation. In the Mitsubishi Gas process the p-tolualdehyde–HBF_4 complex is thermally decomposed in continuous operation in a distillation column and the HF/BF_3 recycled. The p-tolualdehyde is produced in 95% yield.

5.4 Summary

The metal carbonyl-catalysed carbonylations of olefins and organic halides have broad synthetic utility as methods for the synthesis of carboxylic acids and their derivatives. The carbonylation of organic halides in particular proceeds under very mild conditions and provides an attractive alternative to conventional methods for introducing a carboxyl function, such as by reaction with cyanide ion. Although these reactions were initially applied to the manufacture of bulk chemicals, recent attention has been focussed on the synthesis of a variety of fine chemicals. The scope of these reactions has been significantly enhanced by the recent application of phase transfer catalysis techniques to these systems.

With the increasing availability of syn gas one may also expect more attention to be focussed in the future on carbonylations of aromatic substrates. An interesting recent development in this context is the syn gas route for the conversion of toluene to styrene discussed in Chapter 7.

References

1. W. Reppe, *Justus Liebig's Ann. Chem.*, **582**, 1 (1953); W. Reppe and H. Vetter, ibid, **582**, 133 (1953).
2. J. Falbe, *Carbon Monoxide in Organic Synthesis*, Chapter 2, Springer-Verlag, Berlin, 1970.
3. A. Mullen, in *New Syntheses with Carbon Monoxide* (J. Falbe, Ed.), Springer-Verlag, Berlin, 1980, p. 243.
4. P. Pino, F. Piacenti and M. Bianchi, in *Organic Syntheses via Metal Carbonyls*, Vol. 2, (I. Wender and P. Pino, Eds.) Wiley, New York, 1977, p. 233.
5. F. Piacenti, M. Bianchi and R. Lazzaroni, *Chim. Ind.* (Milan) **50**, 318 (1968).
6. L. Cassar, G. P. Chiusoli and F. Guerrieri, *Synthesis*, 509 (1973).
7. J. Tsuji, *Advan. Org. Chem.*, **6**, 109 (1969).
8. K. Bittler, N. V. Kutepow, D. Neubauer and H. Reis, *Angew. Chem.*, **80**, 352 (1968).
9. J. F. Knifton, *J. Org. Chem.*, **41**, 2885 (1976).
10. E. N. Frankel, F. L. Thomas and W. K. Rohwedder, *Advan. Chem. Ser.*, **132**, 145 (1974).
11. H. Bahrmann, in *New Syntheses with Carbon Monoxide* (J. Falbe, Ed.), Springer-Verlag, Berlin, 1980, p. 372.
12. P. Hofmann, K. Kosswig and W. Schaefer, *Ind. Eng. Chem. Prod. Res. Dev.*, **19**, 330 (1980).
13. R. F. Heck and D. S. Breslow, *J. Am. Chem. Soc.*, **85**, 2779 (1963).
14. R. F. Heck, *J. Am. Chem. Soc.*, **85**, 2013 (1963).
15. J. Tsuji, *Acc. Chem. Res.*, **2**, 144 (1969).
16. J. Tsuji and K. Ohno, *Advan. Chem. Ser.*, **70**, 155 (1968).
17. J. Tsuji, M. Morikawa and J. Kiji, *Tetrahedron Letters*, 1061 (1963).
18. J. F. Knifton, *J. Org. Chem.*, **41**, 793 (1976).
19. J. H. Craddock, A. Hershman, F. E. Paulik and J. F. Roth, *US Patent*, 3,816,488 (1974); J. H. Craddock, J. F. Roth, A. Hershman and F. E. Paulik, *US Patent*, 3,989,747 (1976) to Monsanto.

20. M. El-Chahawi, U. Prange, W. Vogt and H. Richtzenhain, *German Patent*, 2,639,327 (1977) to Dynamit Nobel.
21. R. Kummer, H. W. Schneider, F. J. Weiss and O. Lemon, *German Patent*, 2,837,815 (1980) to BASF.
22. *Belgian Patent*, 770,615 (1972) to BASF.
23. J. Tsuji, I. Kiji and S. Hosaka, *Tetrahedron Letters*, 605 (1974).
24. W. E. Billups, W. E. Walker and T. C. Shields, *Chem. Commun.*, 1067 (1971).
25. J. Knifton, *J. Catal.*, **60**, 27 (1979).
26. D. M. Fenton and K. L. Olivier, *Chem. Tech.*, **2**, 220 (1972).
27. G. P. Chiusoli, *Pure Appl. Chem.*, **52**, 635 (1980).
28. D. Medema, R. van Helden and C. F. Kohll, *Inorg. Chim. Acta*, **3**, 255 (1969).
29. D. M. Fenton and P. J. Steinwald, *J. Org. Chem.*, **37**, 2034 (1972).
30. H. S. Kesling, *British Patent*, 2,064,353 (1980) to Atlantic Richfield.
31. G. Consiglio and P. Pino, *Chimia*, **30**, 193 (1976).
32. Y. Iwashita and M. Sakuraba, *Tetrahedron Letters*, 2409 (1971).
33. N. von Kutepow, *German Patent*, 2,445,193 (1976) to BASF.
34. M. El-Chahawi and U. Prange, *Chem. Ztg.*, **102**, 1 (1978).
35. M. El-Chahawi and H. Richtzenhain, *German Patent*, 2,240,398 and 2,240,399 (1974) to Dynamit Nobel.
36. M. El-Chahawi, U. Prange, H. Richtzenhain and W. Vogt, *German Patent*, 2,553,931 (1977) to Dynamit Nobel.
37. K. Redecker, M. El-Chahawi, H. Richtzenhain and W. Vogt, *German Patent*, 2,446,657 (1976) to Dynamit Nobel.
38. M. El-Chahawi, U. Prange, H. Richtzenhain and W. Vogt, *German Patent*, 2,606,655 (1977) to Dynamit Nobel.
39. A. Schoenberg, I. Bartoletti and R. F. Heck, *J. Org. Chem.*, **39**, 3318 (1974).
40. U. Prange, M. El-Chahawi, H. Richtzenhain and W. Vogt, *German Patent*, 2,557,011 (1977) to Dynamit Nobel.
41. M. El-Chahawi, H. Richtzenhain and W. Vogt, *German Patent*, 2,410,782 (1975) to Dynamit Nobel.
42. A. Moro, M. Foa and L. Cassar, *British Patent*, 1,560,609 (1980) to Montedison.
43. B. Falk, W. Vogt and H. Richtzenhain, *German Patent*, 2,704,191 (1978) to Dynamit Nobel.
44. U. Prange, M. El-Chahawi, H. Richtzenhain and W. Vogt, *German Patent*, 2,509,017 (1976) to Dynamit Nobel.
45. *Belgian Patent*, 877,229 (1979) to Dynamit Nobel.
46. M. El-Chahawi and H. Richtzenhain, *German Patent*, 2,259,072 (1974) to Dynamit Nobel.
47. G. P. Chiusoli, *Angew. Chem.*, **72**, 74 (1960); *Chim. Ind.* (Milan) **41**, 503 (1959).
48. F. Montino and M. Caseley, *German Patent*, 1,936,725 (1970) to Montecatini.
49. U. Prange, H. Richtzenhain and W. Vogt, *German Patent*, 2,436,788 (1976) to Dynamit Nobel.
50. U. Prange, M. El-Chahawi, H. Richtzenhain and W. Vogt, *German Patent*, 2,524,389 (1976) to Dynamit Nobel.
51. M. El-Chahawi, U. Prange, H. Richtzenhain and W. Vogt, *German Patent*, 2,403,483 (1975) to Dynamit Nobel.
52. J. P. Collman, S. R. Winter and R. G. Komoto, *J. Am. Chem. Soc.*, **95**, 249 (1973).
53. J. J. Brunet, C. Sidot and P. Caubere, *J. Org. Chem.*, **46**, 3147 (1981); see also B. Loubinoux, B. Fixari, J. J. Brunet and P. Caubere, *J. Organometal. Chem.*, **105**, C22 (1976).

54. J. J. Brunet, C. Sidot and P. Caubere, *J. Organometal. Chem.*, **204**, 229 (1980).
55. J. J. Brunet, C. Sidot, B. Loubinoux and P. Caubere, *J. Org. Chem.*, **44**, 2199 (1979).
56. J. K. Stille and K. S. Y. Lau, *Acc.·Chem. Res.*, **10**, 434 (1977).
57. J. K. Stille and P. K. Wong, J. Org. Chem., **40**, 532 (1975).
58. J. K. Stille, L. F. Hines, R. W. Fries, P. K. Wong, D. E. James and K. Lau, *Advan. Chem. Ser.*, **132**, 90 (1974).
59. T. Yukawa, N. Yamakami, M. Honna, Y. Komachiya and H. Wakamatsu, *German Patent*, 2,364,039 (1974) to Ajinomoto.
60. A. Schoenberg and R. F. Heck, *J. Am. Chem. Soc.*, **96**, 7761 (1974).
61. W. P. Weber and G. W. Gokel, *Phase Transfer Catalysis in Organic Synthesis*, Springer-Verlag, Berlin, 1977.
62. C. M. Starks and C. Liotta, *Phase Transfer Catalysis, Principles and Techniques*, Academic Press, New York, 1978.
63. H. Alper, *Advan. Organometal. Chem.*, **19**, 183 (1981).
64. L. Cassar, in *Fundamental Research in Homogeneous Catalysis* (M. Tsutsui and R. Ugo, Eds.), Plenum Press, New York, 1977, p. 115.
65. H. Alper and H. des Abbayes, *J. Organometal. Chem.*, **134**, C11 (1977).
66. L. Cassar and M. Foa, *J. Organometal. Chem.*, **134**, C15 (1977).
67. L. Cassar and M. Foa, *German Patent*, 2,801,886 (1978) to Montedison.
68. H. des Abbayes and A. Buloup, *Tetrahedron Letters*, **21**, 4393 (1980).
69. H. des Abbayes and A. Buloup, *J. Organometal, Chem.*, **179**, C21 (1979).
70. S. Gambarotta and H. Alper, *J. Organometal, Chem.*, **212**, C23 (1981).
71. S. Gambarotta and H. Alper, *J. Organometal. Chem.*, **194**, C19 (1980).
72. J. J. Brunet, C. Sidot and P. Caubere, *Tetrehedron Letters*, 1013 (1981).
73. L. Cassar, M. Foa and A. Gardano, *J. Organometal. Chem.*, **121**, C55 (1976).
74. L. Cassar, M. Foa and A. Gardano, *US Patent*, 4,034,004 (1977) to Montedison.
75. N. N. Crounse, *Org. React.*, **5**, 290 (1949).
76. W. F. Gresham and G. E. Tabet, *US Patent*, 2,485,237 (1949) to Du Pont.
77. S. Fujiyama and T. Kasahara, *Hydrocarbon Process.*, **57**(11), 147 (1978).

CHAPTER 6

METHANOL AND FORMALDEHYDE

In the preceding two chapters we have discussed the various reactions of lower olefins with carbon monoxide and syn gas that are widely used in the manufacture of industrial chemicals. In this and the next chapter we shall examine the direct transformation of methanol to industrial chemicals. Much of the chemistry on which these processes are based has been known since the 1940s. It was not until quite recently, however, that the rising prices of olefin feedstocks provided an economic incentive for the development of methanol-based processes. This precipitated an intensive research effort which resulted in many process improvements. Some of these methanol-based processes are already displacing existing olefin-based routes. The classic example is the Monsanto process for acetic acid manufacture via methanol carbonylation (see Chapter 7).

Before going on to discuss the reactions of methanol we shall first take a look at the manufacture of methanol from syn gas.

6.1 Methanol Synthesis

Methanol is a basic industrial chemical that is produced on a scale of several million tons per annum. It has been manufactured since the 1920s via the hydrogenation of carbon monoxide over solid catalysts at elevated temperatures and pressures (Reaction 1). The subject has been recently reviewed by Kung [1] and Frohning [2].

$$CO + 2 H_2 \rightleftharpoons CH_3OH \qquad (1)$$

Methanol plants are generally integrated with an ammonia plant because the process and ancillary equipment used are similar. In addition the methanol plant can use the CO_2 produced by the water gas shift reaction in an ammonia synthesis plant. It is treated with methane and steam over a promoted nickel catalyst at 800°C.

$$3 CH_4 + CO_2 + 2 H_2O \longrightarrow 4 CO + 8 H_2 \qquad (2)$$

127

CATALYSTS

The zinc chromium oxide ($ZnO-Cr_2O_3$) catalyst used initially exhibited a low activity and required high pressures (250–750 bar) and temperatures (350–450°C) for economically interesting conversions. Subsequently, more active catalysts based on a zinc copper oxide, supported on alumina or chromia, were developed by ICI. With these catalysts the operating temperature and pressure could be reduced to 50–100 bar and 250–300°C, respectively. Modern methanol plants are exclusively based on this medium pressure process.

The maximum yield of methanol obtainable in Reaction 1 is determined by the position of the thermodynamic equilibrium. Since Reaction 1 is exothermic and methanol formation is accompanied by a decrease in mole number the methanol yield increases with increasing pressure and decreasing temperature. At atmospheric pressure the $\Delta G°$ of the reaction is negative only at temperatures below 140°C. However, temperatures in excess of 140°C are needed with the current catalysts for reasonable activity. This means that the use of relatively high (50–100 bar) pressures are essential for economically viable methanol yields. Another limitation to the zinc copper oxide catalysts currently in use is their sensitivity to sulphur poisoning which means that the syn gas feed must be rigorously purified. There is a definite need, therefore, for a more active, low-temperature catalyst. The pressure could then be reduced to a more convenient range of 5–10 bar.

An important requirement of a methanol synthesis catalyst is an ability to activate carbon monoxide non-dissociatively. Dissociative adsorption, in contrast, leads to hydrocarbon formation (see Chapter 3). The reported [3–5] selective methanol synthesis activity of supported Pd, Pt, Ir and Rh catalysts is consistent with the fact that these metals adsorb CO associatively.

Ichikawa, for example, showed that pyrolysed rhodium carbonyl clusters supported on strongly basic oxides such as ZnO and MgO are selective methanol synthesis catalysts at 220°C and sub-atmospheric pressures [4]. Conversions are, however, relatively low (*ca.* 10%). Interestingly, when certain amphoteric oxides, such as La_2O_3, TiO_2 and ZrO_2, are used as supports, ethanol is the major product. The latter is formed via homologation of the initially formed methanol (see next Chapter). When acidic oxide supports, such as Al_2O_3 and SiO_2, are used, hydrocarbon formation is the predominant reaction.

Novel palladium catalysts, produced by impregnation of Na_2PdCl_4 on Al_2O_3 and SiO_2 have also been shown to be selective methanol synthesis catalysts at sub-atmospheric pressures and 180°C [5]. The activity of these catalysts is enhanced when they are doped with Li or Na cations.

Reaction 1 is also catalysed by soluble complexes of Group VIII metals, e.g. Co [6], rhodium [7] and ruthenium [8–10]. The actual catalysts are the metal carbonyls which are added as such, e.g. $Co_2(CO)_8$, or are formed *in situ* via reduction of the metal salts with syn gas. The relative reactivities appear to be $Rh > Ru > Co$ (direct comparison is difficult due to the different conditions employed).

Cobalt catalysts are generally non-selective and afford complex mixtures consisting of methanol, higher alcohols, formates and glycols [6]. Ruthenium carbonyls, on the other hand, are highly selective [8–10] but require forcing conditions ($> 200°C$ and > 1000 bar). The higher selectivity of Ru compared to Co is a consequence of the much lower acidity of $H_2Ru(CO)_4$ compared to $HCo(CO)_4$. This precludes further homologation of the initially formed methanol (see following chapter).

Bradley [8] studied the hydrogenation of carbon monoxide in tetrahydrofuran at high temperatures and pressures ($> 225°C$ and 1300 bar) in the presence of various ruthenium complexes. Regardless of which catalyst precursor was used infrared analysis after each run revealed the presence only of $Ru(CO)_5$. The only products observed were methanol and methyl formate in a molar ratio of $4 : 1$. The addition of triphenylphosphine had a beneficial effect on methanol formation, a selectivity of $> 95\%$ being observed with a $1 : 1$ CO/H_2 mixture at $275°C$ and 1200 bar in the presence of the phosphine.

Rhodium carbonyls also catalyse methanol synthesis. In the reported studies [7], however, emphasis was placed on ethylene glycol formation (see Chapter 9) and very high pressures ($500–1500$ bar) were used to achieve this goal. Although methanol formation is favoured at lower pressures the optimum conditions for methanol synthesis have not been clearly defined.

The studies discussed above have demonstrated the feasibility of homogeneous catalysis of methanol synthesis from syn gas. However, because of the extreme conditions of temperature and pressures required, these systems are less attractive than the existing heterogeneous catalysts and the goal of finding an active low pressure catalyst has yet to be achieved. Since rhodium carbonyls generally exhibit superior activities they would seem to warrant further study as methanol synthesis catalysts.

MECHANISM

Basically two mechanisms have been considered to explain the formation of methanol over heterogeneous oxide catalysts [1]. In the first mechanism (Scheme 6.1) adsorbed CO undergoes a sequence of hydrogenation steps involving intermediates bonded to the surface through the carbon atom.

In the second mechanism (Scheme 6.2) the initial step is insertion of CO into a surface hydroxyl to form a surface formate. This is followed by subsequent hydrogenation and dehydration to give a surface methoxide and then methanol.

In this mechanism reaction intermediates are bonded to the surface through

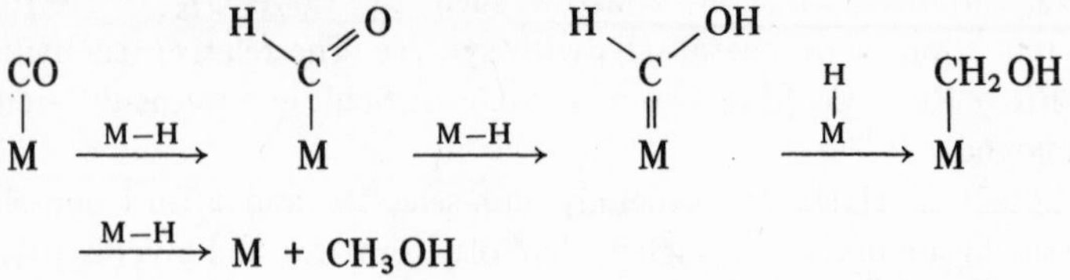

$$\underset{\text{M}}{\overset{\text{CO}}{|}} \xrightarrow{\text{M--H}} \underset{\text{M}}{\overset{\text{H}}{\underset{|}{\overset{}{\text{C}}}}}{\overset{}{\diagup O}} \xrightarrow{\text{M--H}} \underset{\text{M}}{\overset{\text{H}}{\underset{\|}{\overset{}{\text{C}}}}}{\overset{}{\diagup OH}} \xrightarrow[\text{M}]{\text{H}} \underset{\text{M}}{\overset{\text{CH}_2\text{OH}}{|}}$$

$$\xrightarrow{\text{M--H}} \text{M} + \text{CH}_3\text{OH}$$

Scheme 6.1

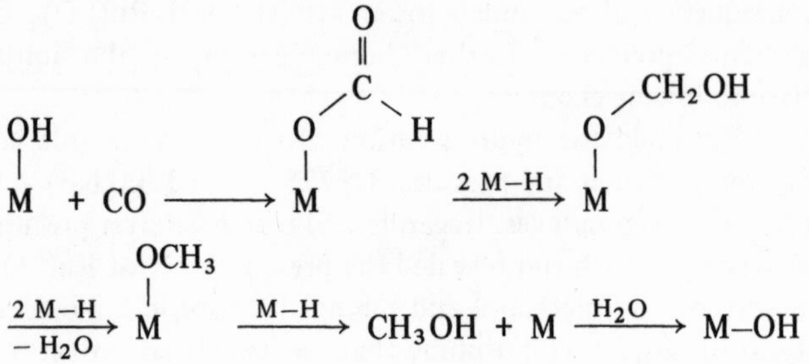

$$\underset{\text{M}}{\overset{\text{OH}}{|}} + \text{CO} \longrightarrow \underset{\text{M}}{\overset{O\overset{\|}{\underset{|}{C}}\diagdown H}{|}} \xrightarrow{\text{2 M--H}} \underset{\text{M}}{\overset{O\diagdown CH_2OH}{|}}$$

$$\xrightarrow[-\,\text{H}_2\text{O}]{\text{2 M--H}} \underset{\text{M}}{\overset{\text{OCH}_3}{|}} \xrightarrow{\text{M--H}} \text{CH}_3\text{OH} + \text{M} \xrightarrow{\text{H}_2\text{O}} \text{M--OH}$$

Scheme 6.2

oxygen. Support for this mechanism derives from the fact that CO is known to react with strongly basic hydroxides, e.g. NaOH, to give formate ion, and that methanol synthesis catalysts contain a strongly basic component (ZnO). In addition, the second component, copper, is known to be an effective catalyst for the hydrogenation of formate esters to methanol.

An attractive feature of the first mechanism is that it is applicable to both the heterogeneous and homogeneous systems. As discussed in Chapter 2 there is ample evidence in favour of a mechanism involving sequential hydrogenation, and formaldehyde as a putative intermediate, for the hydrogenation of CO to oxygenated products. The concurrent formation of methanol and ethylene glycol in the homogeneous systems is thought to be a consequence of competing hydrogenation of, and CO insertion into, the $\text{M--CH}_2\text{OH}$ intermediate.

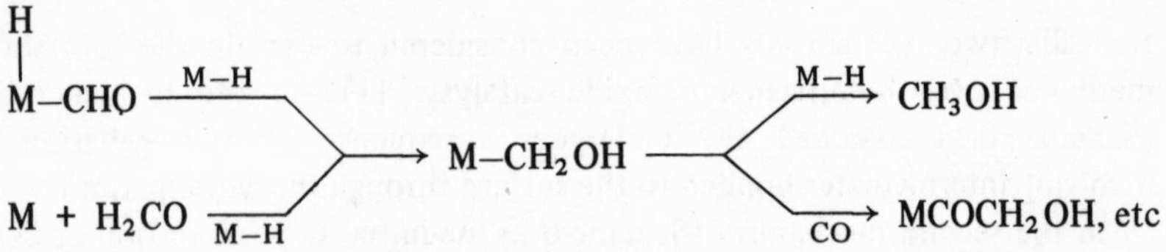

$$\begin{array}{ccc} \overset{\text{H}}{\underset{|}{\text{M--CHO}}} & \xrightarrow{\text{M--H}} & \\ & \searrow & \\ & \text{M--CH}_2\text{OH} & \\ & \nearrow & \\ \text{M} + \text{H}_2\text{CO} & \xrightarrow{\text{M--H}} & \end{array} \begin{array}{c} \xrightarrow{\text{M--H}} \text{CH}_3\text{OH} \\ \\ \xrightarrow{\text{CO}} \text{MCOCH}_2\text{OH, etc} \end{array}$$

Scheme 6.3

Definitive mechanistic study of methanol synthesis has generally been hampered by the high pressure requirement and it is not possible, on the basis of the available evidence, to distinguish unequivocally between the various mechanisms.

6.2 Formaldehyde Synthesis

Methanol is widely used as an industrial solvent and as a raw material for a variety of industrial chemicals such as dimethyl terephthalate, methyl methacrylate, methyl chloride, methylamines and acetic acid. By far the most important use is, however, in the manufacture of formaldehyde which accounts for 40–50% of methanol consumption. Formaldehyde is in turn an important raw material for a range of formaldehyde resins, such as phenol-formaldehyde, urea-formaldehyde and polyacetal resins. Moreover, formaldehyde could in the future become an important intermediate in the manufacture of several commodity chemicals from methanol (see below).

The conversion of methanol to formaldehyde by air oxidation (Reaction 3) is well established commercial technology. Most of the formaldehyde (*ca.* 80%)

$$CH_3OH + \tfrac{1}{2} O_2 \longrightarrow H_2CO + H_2O \tag{3}$$

is manufactured by the BASF process [11] which employs a silver catalyst at 650–700°C and atmospheric pressure and gives a 91% selectivity at 99% methanol conversion. The rest is manufactured by the FORMOX process [12] which employs a ferric molybdate catalyst at 300°C and atmospheric pressure, and affords 97% selectivity at > 99% methanol conversion.

Mitsubishi Gas [13] has developed a methanol dehydrogenation process (Reaction 4) which uses a Cu/Zn/Se catalyst at 600°C, in the absence of air, and gives formaldehyde in 92% selectivity at 60% methanol conversion. In this process the methanol conversion level is subject to thermodynamic limitations.

$$CH_3OH \rightleftharpoons H_2CO + H_2 \tag{4}$$

This technology has not yet been commercialized, but has the advantage that it produces essentially water-free formaldehyde together with valuable hydrogen as a co-product. It could become attractive in the future, especially if the methanol-formaldehyde mixture formed could be used directly without further purification.

6.3 Formaldehyde Chemistry

About 60% of the formaldehyde currently manufactured is consumed as a raw material for formaldehyde resins (see Table 6.I).

TABLE 6.I
Formaldehyde consumption

Product	% of Total
Urea/formaldehyde resins	25
Phenol/formaldehyde resins	24
Butane diol	8
Acetal resins	7
Pentaerythritol	6
Hexamethylenetetramine .	4
Melamine/formaldehyde resins	4
Urea/formaldehyde concentrates	4
Chelating agents	4
Miscellaneous	14

In the future, as syn gas-derived methanol becomes an increasingly important base chemical, formaldehyde could become a key intermediate to various bulk chemicals that are presently produced from lower olefins. We shall now examine a few of these possibilities.

ETHYLENE GLYCOL SYNTHESIS VIA FORMALDEHYDE CARBONYLATION

The Du Pont process for the manufacture of ethylene glycol from formaldehyde proceeds by the following steps [14].

$$H_2CO + CO + H_2O \xrightarrow{H_2SO_4} HOCH_2CO_2H \tag{5}$$

$$HOCH_2CO_2H + MeOH \longrightarrow HOCH_2CO_2Me + H_2O \tag{6}$$

$$HOCH_2CO_2Me + 2 H_2 \longrightarrow HOCH_2CH_2OH + CH_3OH \tag{7}$$

When this is coupled with methanol manufacture (Reaction 1) and oxidation to formaldehyde (Reaction 3) this leads to the overall conversion:

$$4 H_2 + 2 CO + \tfrac{1}{2} O_2 \longrightarrow HOCH_2CH_2OH + H_2O \tag{8}$$

The Du Point plant was operated in the United States from 1940 to 1968 when it was closed down because of competition from the more economical ethylene-

$$CH_2{=}CH_2 + \tfrac{1}{2} O_2 \xrightarrow[ca.\ 250°C]{Ag/Al_2O_3} \overset{O}{\overset{\triangle}{CH_2{-}CH_2}} \tag{9}$$

$$\overset{O}{\overset{\triangle}{CH_2{-}CH_2}} + H_2O \longrightarrow HOCH_2CH_2OH \tag{10}$$

based process. The latter involves oxidation of ethylene to ethylene oxide, over a silver catalyst at *ca*. 250°C, followed by hydrolysis (Reactions 9 and 10).

However, recent price trends in petrochemical feedstocks have led to a re-evaluation of syn gas *vs.* ethylene-based routes to ethylene glycol and, hence, to a revival of interest in variations of the Du Pont process. A recent variant, patented by Chevron [15], involves the carbonylation of formaldehyde with CO/H_2 (the H_2 does not react) in the presence of HF as catalyst and solvent. The hydroxyacetic acid so produced is esterified with ethylene glycol and the hydrogen removed in the carbonylation step is then used to hydrogenate the hydroxyacetic acid ester. This obviates the need for the methanol recycle demanded by the Du Pont process and allows for the use of syn gas instead of H_2 and CO separately.

$$H_2CO + CO + H_2O \xrightarrow{\text{HF}} HOCH_2CO_2H \tag{11}$$

$$HOCH_2CO_2H + HOCH_2CH_2OH \longrightarrow HOCH_2CO_2CH_2CH_2OH + H_2O \tag{12}$$

$$HOCH_2CO_2CH_2CH_2OH + H_2 \xrightarrow{\text{catalyst}} 2 HOCH_2CH_2OH \tag{13}$$

Reaction 11 is also catalysed by rhodium compounds in the presence of iodide promotors [16]. A plausible mechanism [16], that has many features in common with the widely accepted mechanism for rhodium/iodide-catalysed carbonylation of methanol (see next chapter), is shown in Scheme 6.4.

$$CH_2O + HI \rightleftharpoons HOCH_2I \xrightarrow{Rh^I} I-Rh^{III}-CH_2OH$$

$$\xrightarrow{CO} I-Rh^{III}-COCH_2OH \xrightarrow{H_2O}$$

$$HOCH_2CO_2H + HI + Rh^I$$

Scheme 6.4

Another route to ethylene glycol involves the hydrocarbonylation of formaldehyde to hydroxyacetaldehyde using Co [17] or Rh [18] catalysts, followed by hydrogenation.

$$H_2CO + CO + H_2 \longrightarrow HOCH_2CHO \tag{14}$$

$$HOCH_2CHO + H_2 \longrightarrow HOCH_2CH_2OH \tag{15}$$

These formaldehyde-based routes to ethylene glycol obviously have commercial potential and have to be compared with the direct syn gas to ethylene glycol route discussed in the next chapter. Ethylene glycol can also be prepared via self-condensation of formaldehyde to hydroxyacetaldehyde (see below).

OTHER CARBONYLATION REACTIONS

In principle, malonic acid (esters) should be available from the alkoxycarbonylation of formaldehyde in the presence of transition metal catalysts such as Co and Rh:

$$H_2CO + 2\,ROH + 2\,CO \longrightarrow CH_2(CO_2R)_2 + H_2O \qquad (16)$$

$$R = H, \text{ alkyl}$$

Bis-carbonylation of formaldehyde dimethylacetal, for example, would afford dimethyl malonate:

$$H_2C(OCH_3)_2 + 2\,CO \longrightarrow CH_2(CO_2CH_3)_2 \qquad (17)$$

To our knowledge, such reactions have not yet been reported but by analogy with the carbonylation of ethers to esters (see Chapter 7) Reaction 17 should be feasible.

The cobalt-catalysed carbonylation of formaldehyde in the presence of acetamide results in the formation of acetylglycine (Reaction 18).

$$H_2CO + CH_3CONH_2 + CO \longrightarrow CH_3CONHCH_2CO_2H \quad (18)$$

This and analogous syntheses of amino acid derivatives are discussed in more detail in Chapter 8.

The selective homologation of formaldehyde to acetaldehyde by reaction with syn gas (Reaction 19) in the presence of a ruthenium carbonyl catalyst and a halide promotor has recently been described [19]. The acetaldehyde is hydrogenated to ethanol in a second stage. This reaction constitutes an alternative to methanol homologation (see next chapter) for acetaldehyde and/or ethanol synthesis. Indeed, Reaction 19 may proceed via the intermediacy of methanol.

$$H_2CO + CO + 2\,H_2 \xrightarrow[75-250°C]{[Ru_3(CO)_{12}-HBr]} CH_3CHO + H_2O \qquad (19)$$

FORMALDEHYDE CONDENSATION REACTIONS

It is well-known that formaldehyde undergoes base- and acid-catalysed condensation reactions (aldol condensations). The most well-known example, from an industrial point of view, is the condensation with acetaldehyde which is used for the industrial manufacture of pentaerythritol:

$$CH_3CHO + 4\,H_2CO \xrightarrow{NaOH} C(CH_2OH)_4 + HCO_2Na \qquad (20)$$

In many cases the initially formed aldol condensation product undergoes dehydration to form unsaturated compounds, following the general equation:

$$RCH_2X + H_2CO \longrightarrow RC(X)=CH_2 + H_2O \tag{21}$$

$X = CO_2R', COR', CN, etc$

This reaction can be used for the synthesis of several important industrial chemicals. For example, Reactions 22–26 were carried out in the presence of a KOH/SiO$_2$ [20] or supported rare earth oxide [21] catalysts.

$$CH_3CO_2CH_3 + H_2CO \xrightarrow[480-490^\circ C]{La_2O_3/SiO_2} CH_2=CHCO_2CH_3 + H_2O \tag{22}$$

$$CH_3CH_2CO_2CH_3 + H_2CO \xrightarrow[440^\circ C]{KOH/SiO_2} CH_2=C(CH_3)CO_2CH_3 + H_2O \tag{23}$$

$$CH_3CH_2CO_2H + H_2CO \xrightarrow[430^\circ C]{KOH/SiO_2} CH_2=C(CH_3)CO_2H + H_2O \tag{24}$$

$$CH_3COCH_3 + H_2CO \xrightarrow[475^\circ C]{La_2O_3/SiO_2} CH_2=CHCOCH_3 + H_2O \tag{25}$$

$$CH_3CH_2CN + H_2CO \xrightarrow[475^\circ C]{La_2O_3/SiO_2} CH_2=C(CH_3)CN + H_2O \tag{26}$$

Selectivities in the range 80–90% were observed at formaldehyde conversions of 30–95%. In principle, it should be possible to carry out this type of reaction with *in situ* generation of formaldehyde via methanol dehydrogenation. Indeed, one example of such a reaction has recently been reported. Thus when a mixture of methanol and methyl propionate was oxidized with air over a CuO/ZnO/TeO$_2$ catalyst at 420°C this resulted in the formation of methyl methacrylate in 81% selectivity at 13% conversion [22].

$$CH_3CH_2CO_2CH_3 + CH_3OH + \tfrac{1}{2}O_2$$

$$\xrightarrow[430^\circ C]{CuO/ZnO/TeO_2} CH_2=C(CH_3)CO_2CH_3 + 2H_2O \tag{27}$$

The self-condensation of formaldehyde in the presence of basic catalysts, such as NaOH, leads to the formation of hydroxyacetaldehyde (Reaction 28). The latter cannot generally be isolated as it undergoes rapid further condensation with formaldehyde to afford carbohydrates. This transformation is usually referred to as the Formose reaction [23].

$$H_2CO + H_2CO \xrightarrow{NaOH} HOCH_2CHO \tag{28}$$

The reaction can be stopped at the hydroxyacetaldehyde stage by using a shape selective, NaOH/zeolite catalyst. For example, condensation of formaldehyde over a NaOH/Na mordenite catalyst at 94°C and 1 bar affords hydroxyacetaldehyde in high ($> 80\%$) selectivity [24]. The remainder consisted of ethylene glycol and sodium formate, formed through the Cannizzaro reaction of hydroxyacetaldehyde with formaldehyde (Reaction 29). No products higher than C_2 were observed.

$$HOCH_2CHO + H_2CO + NaOH \longrightarrow HOCH_2CH_2OH + HCO_2Na \qquad (29)$$

Reaction 28, followed by catalytic hydrogenation of the hydroxyacetaldehyde product, has commercial potential as a formaldehyde-based route to ethylene glycol. This route is preferable to conversion by Reaction 29 since the latter consumes stoichiometric quantities of H_2CO and NaOH.

In principle, combination of dehydrogenation, basic and shape selective properties into a multifunctional catalyst could provide a method for the direct conversion of methanol to ethylene glycol via the sequence:

$$CH_3OH \longrightarrow H_2CO + H_2 \qquad (30)$$

$$H_2CO + H_2CO \longrightarrow HOCH_2CHO \qquad (31)$$

$$HOCH_2CHO + H_2 \longrightarrow HOCH_2CH_2OH \qquad (32)$$

Net reaction: $2\ CH_3OH \longrightarrow HOCH_2CH_2OH + H_2 \qquad (33)$

For example, copper chromite or a ruthenium compound in combination with a base and a shape selective zeolite may be expected to catalyse Reaction 33. To our knowledge such a transformation, which has obvious commercial potential, has not yet been described.

ADDITION OF FORMALDEHYDE TO OLEFINS

The ene addition of formaldehyde to olefins in the presence of an acid catalyst or at elevated temperatures, is the well-known Prins reaction [25].

$$RCH{=}CH_2 + H_2CO \longrightarrow RCH{=}CHCH_2OH \qquad (34)$$

Terminal alcohols can be prepared by subsequent hydrogenation of the product. These reactions have limited commercial application, however, since the same alcohols are available from olefin hydroformylation (see Chapter 4). An interesting application of the Prins reaction is the recently reported HF-catalysed addition of formaldehyde to allyl alcohol to give 4-hydroxybutyraldehyde [26].

$$CH_2{=}CHCH_2OH + H_2CO \xrightarrow{\text{HF}} HO(CH_2)_3CHO \qquad (35)$$

Hydrogenation of the latter affords the industrially important monomer, 1,4-butanediol. Traditionally 1,4-butanediol has been manufactured by condensation of two moles of formaldehyde with acetylene (see Chapter 4 for an alternative synthesis by hydroformylation).

6.4 Methyl Formate

When the dehydrogenation of methanol is performed over a copper catalyst modified with a Group IVA element (Zr, Ti) or a lanthanide the result is the formation of methyl formate in up to 97% selectivity at *ca.* 20–40% methanol conversion [27].

$$2 \; CH_3OH \; \xrightarrow[\textit{ca. 200°C}]{[Cu/Zr \; carbonates]} \; HCO_2CH_3 \; + \; H_2 \qquad (36)$$

The mechanism of the reaction is not known but may involve the intermediacy of formaldehyde as shown:

$$H_2CO \; + \; CH_3OH \; \rightleftharpoons \; H_2C(OCH_3)OH \; \longrightarrow \; HCO_2CH_3 \; + \; H_2 \qquad (37)$$

Methyl formate can also be produced from the reaction of methanol with carbon monoxide under pressure, in the presence of a basic catalyst [28].

$$CH_3OH \; + \; CO \; \longrightarrow \; HCO_2CH_3 \qquad (38)$$

Thus, methyl formate is an example of a C_2 oxygenate that is readily synthesized from syn gas/methanol. Applications of methyl formate are, however limited. It is mainly converted to dimethylformamide by reaction with dimethylamine.

$$HCO_2CH_3 \; + \; (CH_3)_2NH \; \longrightarrow \; HCON(CH_3)_2 \; + \; CH_3OH \qquad (39)$$

It can also be used as a source of formic acid via autocatalytic hydrolysis [28–31].

$$HCO_2CH_3 \; + \; H_2O \; \longrightarrow \; HCO_2H \; + \; CH_3OH \qquad (40)$$

Traditionally, formic acid has been recovered as a by-product from the manufacture of acetic acid by the liquid phase oxidation of *n*-butane [29, 30]. With the advent of the more economical low pressure methanol carbonylation process for acetic acid (see next chapter) this source of formic acid is gradually being phased out. This has led to renewed interest in formic acid manufacture from methyl formate [31]. Since the methanol consumed in Reaction 38 is regenerated in Reaction 40 the overall process in effect converts CO and water to formic acid.

Another interesting transformation of methyl formate is its isomerisation to acetic acid which reportedly occurs in 95% yield in the presence of CO, a nickel catalyst and an iodide promotor at 180°C [32].

$$HCO_2CH_3 \xrightarrow[180^\circ C]{Ni/MeI/CO} CH_3CO_2H \qquad (41)$$

This provides a two-step process for the conversion of methanol to acetic acid. It is not readily apparent, however, that such an alternative process offers advantages over the direct carbonylation of methanol (see next chapter). Reaction 41 is also catalysed by palladium [33] and rhodium [34] compounds in the presence of iodide promotors. It can also be extended to higher esters of formic acid, e.g. ethyl formate is isomerised to propionic acid [33].

6.5 Summary

In the context of chemicals manufacture from syn gas/methanol, formaldehyde constitutes a possible key intermediate to various downstream derivatives. Several of these possibilities are summarised schematically in Figure 6.1.

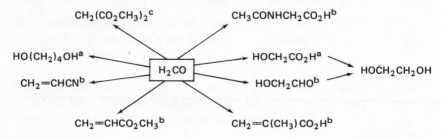

^a Commercial process
^b Demonstrated but not commercialized
^c Speculative

Figure 6.1. Formaldehyde as a key intermediate.

Another product that merits further attention as a possible key intermediate to downstream derivates is methyl formate.

References

1. H. H. Kung, *Catal. Rev.*, **22**, 235 (1980).
2. C. D. Frohning, in *New Syntheses with Carbon Monoxide* (J. Falbe, Ed.), Springer-Verlag, Berlin, 1980, p. 309.
3. M. L. Poutsma, L. F. Elek, D. A. Ibarbia, A. P. Risch and J. A. Rabo, *J. Catal.*, **52**, 157 (1978).
4. M. Ichikawa, *Bull. Chem. Soc.* Japan, **51**, 2268 (1978); *J. Chem. Soc. Chem. Commun.*, 566 (1978); M. Ichikawa, K. Sekizawa, K. Shikakura and M. Kawai, *J. Mol. Catal.*, **11**, 167 (1981).

5.   Y. Kikuzono, S. Kagami, S. Naito, T. Onishi and K. Tamaru, *Chem. Letters*, 1249 (1981).
6.   J. W. Rathke and H. M. Feder, *J. Am. Chem. Soc.*, **100**, 3623 (1978).
7.   R. L. Pruett and W. F. Walker, *US Patent* 3,833,634 (1974); L. Kaplan, *US Patent* 4,162,261 (1979) both to Union Carbide.
8.   J. S. Bradley, *J. Am. Chem. Soc.*, **101**, 7419 (1979); *Fundamental Research in Homogeneous Catalysis* (M. Tsutsui, Ed.), Vol. 3, Plenum Press, New York, 1979, p. 165.
9.   R. B. King, A. D. King and K. Tanaka, *J. Mol. Catal.*, **10**, 75 (1980).
10. B. D. Dombek, *J. Am. Chem. Soc.*, **102**, 6855 (1980).
11. G. A. Halbritter, W. Muelthaler, H. Sperber, H. Diem, C. Dudeck and G. Lehman, *US Patent* 4,072,717 (1968) to BASF.
12. P. Courty, *US Patent* 3,716,497 (1973) to Institut Français du Pétrole.
13. M. Osugi and T. Uchiyama, *US Patent* 4,054,609 (1977) to Mitsubishi Gas.
14. W. F. Gresham, *US Patent* 2,636,046 (1948) to Du Pont.
15. S. Suzuki, *US Patent* 4,087,470 (1976) to Chevron.
16. S. Lapporte and V. P. Kurkov, in *Organotransition-Metal Chemistry* (Y. Ishii and M. Tsutsui, Eds.), Plenum Press, New York, 1975, p. 199.
17. T. Yukawa and H. Wakamatsu, *British Patent* 1,408,857 (1975) to Ajinomoto.
18. A. Spencer, *Europ. Patent Appl.* 2,908 (1978) to Monsanto; R. W. Goetz, *US Patent* 4,200,765 (1980) to National Distillers.
19. D. W. Smith, *US Patent* 4,267,384 (1981) to National Distillers.
20. A. J. C. Pearson, *US Patent* 3,840,587 and 3,840,588 (1974) to Monsanto.
21. T. C. Shapp, A. E. Blood and H. J. Hagemeyer, *US Patent* 3,578,702 (1971) and 3,701,798 (1972) to Eastman Kodak.
22. F. Merger and G. Fouquet, *German Patent* 3,004,467 (1981) to BASF.
23. R. F. Socha, A. H. Weiss and M. M. Sakharov, *J. Catal.*, **67**, 207 (1981); T. I. Khomenko, M. M. Sakharov and O. A. Golovina, *Russ. Chem. Rev.*, **49**, 570 (1980).
24. A. H. Weiss, S. Trigerman, G. Dunnells, V. A. Likhubov and E. Biron, *Ind. Eng. Chem. Process Des. Dev.*, **18**, 522 (1979).
25. D. R. Adams and S. P. Bhatnagar, *Synthesis*, 661 (1977).
26. S. Suzuki, *US Patent* 4,029,711 (1977) to Chevron.
27. M. Yoneoka, *German Patent* 2,716, 842 (1977) to Mitsubishi Gas.
28. *Hydrocarbon Process.*, **56** (11), 166 (1977).
29. A. Aguilo and T. Horlenko, *Hydrocarbon Process.*, **59** (11) 120 (1980).
30. M. P. Czaikowski and A. R. Bayne, *Hydrocarbon Process.*, **59** (11), 103 (1980).
31. A. Peltzman, *Oil Gas J.*, Nov. 16, 1981, p. 103.
32. *Japanese Patent* 81 73,040 (1981) to Mitsubishi Gas; *CA*, **95**, 149967q (1981).
33. *Japanese Patent* 81 22,745 (1981) to Mitsubishi Gas; *CA*, **95**, 80169m (1981).
34. F. J. Bryant, W. R. Johnson and T. C. Singleton, A. C. S. Meeting, Dallas, 1973, *General Papers Petrochemicals*, p. 193.

CHAPTER 7

METHANOL CARBONYLATION AND RELATED CHEMISTRY

Some of the most interesting syn gas-based chemistry to be developed in recent years involves the insertion of carbon monoxide into C–O bonds as a key feature. This technology encompasses a plethora of industrially important transformations. Some of these processes have already found application and others stand at the threshold of commercialization. Indeed, it is in this area that many of the fundamental changes in petrochemical technology are currently taking place. These reactions can be broadly described by the two general equations:

$$R-O-R' + CO \longrightarrow R-\overset{\overset{\displaystyle O}{\|}}{C}-OR' \qquad (1)$$

$$R-O-R' + CO + 2\,H_2 \longrightarrow RCH_2OR' + H_2O \qquad (2)$$

R = alkyl
R' = H, alkyl, acyl

Reaction 1 involves the simple insertion of carbon monoxide into a C–O bond. When the substrate is an alcohol (R' = H) this results in the formation of a carboxylic acid. With ethers (R' = alkyl) or esters (R' = acyl) this results in the formation of esters and anhydrides, respectively. Reaction 2 can be described as a *hydrocarbonylation* as it involves reaction of both CO and H_2. The reaction is often referred to as *homologation* since it results in the extension of the carbon chain of R by one methylene unit.

Reactions 1 and 2 are catalysed by Group VIII transition metals, notably Co, Rh and Ni, usually in the presence of halide promotors. All of these transformations involve the following fundamental steps:

$$ROR' + HX \longrightarrow RX + R'OH \qquad (3)$$

$$RX + M \longrightarrow R-M-X \qquad (4)$$

$$R-M-X + CO \longrightarrow RCO-M-X \qquad (5)$$

The fate of the acylmetal intermediate depends on whether or not hydrogen is present. In its absence nucleophilic attack of R'OH occurs:

$$RCO-M-X + R'OH \longrightarrow RCO_2R' + M + HX \qquad (6)$$

In the presence of hydrogen the acylmetal species undergoes hydrogenolysis to give an aldehyde and, subsequently, an alcohol (Reaction 7 and 8). The overall result is then homologation of the R group.

$$RCO-M-X + H_2 \longrightarrow RCHO + M + HX \qquad (7)$$

$$RCHO + H_2 \xrightarrow{[M]} RCH_2OH \qquad (8)$$

The most well-known example of this important group of reactions is without doubt methanol carbonylation to acetic acid.

7.1 Acetic Acid via Methanol Carbonylation

Acetic acid and acetic anhydride are major industrial chemicals used in the manufacture of, *inter alia*, vinyl acetate and cellulose acetate. Acetic acid and various acetate esters are also important industrial ·solvents. Up until quite recently the industrial manufacture of acetic acid was dominated by two processes: *n*-butane oxidation and ethylene oxidation [1]. In the Celanese process *n*-butane is oxidized with air at elevated temperatures and pressures, in the presence of a cobalt catalyst, to give acetic acid in *ca.* 50% yield.

$$CH_3CH_2CH_2CH_3 + 2\tfrac{1}{2}O_2 \xrightarrow[\substack{150-225°C \\ 55\,bar}]{[Co(OAc)_2]} 2\,CH_3CO_2H + H_2O \qquad (9)$$

In the Wacker process ethylene is oxidized with air in the presence of a $PdCl_2/CuCl_2$ catalyst to give acetaldehyde. The latter is subsequently oxidized with air to acetic acid. Both steps proceed in *ca.* 95% yield.

$$CH_2{=}CH_2 + \tfrac{1}{2}O_2 \xrightarrow[125°C/10\,bar]{[PdCl_2/CuCl_2]} CH_3CHO \qquad (10)$$

$$CH_3CHO + \tfrac{1}{2}O_2 \xrightarrow[60°C/1\,bar]{[Mn(OAc)_2]} CH_3CO_2H \qquad (11)$$

Both of these processes use feedstocks, *n*-butane and ethylene, that are derived from naphtha (or ethane) cracking. In recent years the rapidly escalating prices of hydrocarbon feedstocks have turned the tables in favour of methanol carbonylation (Reaction 12) as the most economical route to acetic acid.

$$CH_3OH + CO \longrightarrow CH_3CO_2H \qquad (12)$$

In his pioneering work on metal-catalysed carbonylations, Reppe [2] showed that the carbonyls of iron, cobalt and nickel are able to catalyse Reaction 12, at temperatures of 250–270°C and pressures of 200–350 bar in the presence of halide promotors. The order of catalytic reactivity is Ni>Co>Fe and the promotor efficiency I>Br>Cl. This finding was developed into a commercial process for acetic acid manufacture by BASF. The process employs a cobalt catalyst and an iodide promotor at *ca.* 200°C and 700 bar [3]. The active catalyst is $HCo(CO)_4$, generated via the reactions:

$$2 CoI_2 + 2 H_2O + 10 CO \longrightarrow Co_2(CO)_8 + 4 HI + 2 CO_2 \qquad (13)$$

$$Co_2(CO)_8 + H_2O + CO \longrightarrow 2 HCo(CO)_4 + CO_2 \qquad (14)$$

The hydrogen iodide reacts with methanol to form methyl iodide (Reaction 19). The latter undergoes oxidative addition to $HCo(CO)_4$ to form, after elimination of HI, a methylcobalt carbonyl (Reaction 15). This is then followed by CO insertion and hydrolysis to give acetic acid and regenerate the catalyst.

$$HCo(CO)_4 + CH_3I \longrightarrow CH_3Co(CO)_4 + HI \qquad (15)$$

$$CH_3Co(CO)_4 + CO \longrightarrow CH_3COCo(CO)_4 \qquad (16)$$

$$CH_3COCo(CO)_4 + H_2O \longrightarrow CH_3CO_2H + HCo(CO)_4 \qquad (17)$$

The severe reaction conditions required for this process constitute a serious drawback. An important technological breakthrough in this area was, therefore, the discovery by Monsanto workers [4, 5] that rhodium compounds in the presence of an iodide promotor are able to catalyse Reaction 12 effectively at atmospheric pressure and 100°C. Acetic acid is produced in essentially quantitative (> 99%) yield, in contrast to the cobalt-catalysed process which gives a 90% yield.

This low pressure carbonylation process was commercialized in 1970 by Monsanto and is currently the most important source of acetic acid accounting for $> 10^6$ tons annual production. The commercial process [6, 7] operates at *ca.* 180°C and 30–40 bar and about 10^{-3} molar catalyst concentration in aqueous acetic acid. The rhodium catalyst may be almost any soluble rhodium compound but is usually added as $RhCl_3 \cdot 3 H_2O$. Various iodine compounds can be used as co-catalysts but methyl iodide or HI are generally used. Both rhodium and iodine are sufficiently expensive that virtually quantitative recycle is essential for economic reasons. Moreover, as the Rh/HI system is highly corrosive much of the equipment has to be constructed from expensive alloys. A further feature of this process is that even in the presence of relatively large amounts of hydrogen no by-product formation is observed, in contrast to the cobalt-catalysed process.

The kinetics of the cobalt- and rhodium-catalysed reactions are significantly different. The former is strongly dependent on both the carbon monoxide pressure and methanol concentration whilst the latter is not. This suggests that different rate-limiting steps are involved and explains the higher methanol conversions and lower operating pressure characteristic of the rhodium system.

The mechanism of the rhodium-catalysed process has been clarified by Forster [8–10]. The active catalyst is $[Rh(CO)_2I_2]^-$ and the catalytic cycle begins with oxidative addition of methyl iodide to this complex (see Scheme 7.1). Evidently the anionic rhodium(I) complex is a powerful nucleophile and this step, although thought to be rate-limiting, is unusually rapid. The resulting methylrhodium(III) complex (I) is kinetically unstable and rapidly isomerises to the acetylrhodium(III) complex (II). The latter reacts with CO to form a labile six-coordinate complex (III) which, in the absence of methanol or water, undergoes reductive elimination to produce acetyl iodide and regenerate the catalyst.

$$[Rh(CO)_2I_2]^- + CH_3I \longrightarrow [CH_3Rh(CO)_2I_3]^-$$
$$(I)$$

$$\longrightarrow [CH_3CORh(CO)I_3]^- \xrightarrow{CO} [CH_3CORh(CO)_2I_3]^-$$
$$(II) \qquad\qquad\qquad (III)$$

$$\longrightarrow CH_3COI + [Rh(CO)_2I_2]^-$$

Scheme 7.1

The catalytic cycle is then completed via Reactions 18 and 19 which produce acetic acid and regenerate the methyl iodide, respectively.

$$CH_3COI + H_2O \longrightarrow CH_3CO_2H + HI \qquad (18)$$

$$CH_3OH + HI \longrightarrow CH_3I + H_2O \qquad (19)$$

Alternatively, in the commercial reaction system, acetic acid may be formed via direct hydrolysis of (III):

$$[CH_3CORh(CO)_2I_3]^- + H_2O \longrightarrow CH_3CO_2H + [Rh(CO)_2I_2]^- + HI \qquad (20)$$
$$(III)$$

When the reaction is carried out at low water concentrations the major product is methyl acetate resulting from methanolysis of acetyl iodide and/or (III).

$$[CH_3CORh(CO)_2I_3]^- + CH_3OH$$
$$\longrightarrow CH_3CO_2CH_3 + [Rh(CO)_2I_2]^- + HI \qquad (21)$$

Despite the enormous commercial success of the Monsanto process industrial research in this area continues unabated. Apparently there is still sufficient incentive to find cheaper and less corrosive catalyst systems that are able to catalyse Reaction 12 under relatively mild conditions. Halcon workers, for example, have reported [11] that the carbonylation of methanol is catalysed by a nickel acetate/tetraphenyltin/methyl iodide system under relatively mild conditions (35 bar and 150°C). Similarly, Union Carbide workers showed [12] that $Mn_2(CO)_{10}/CH_3I$ catalyses the carbonylation of methanol to methyl acetate in a water-free system at 120°C and 100 bar. The commercial viability of these alternative systems remains to be demonstrated.

The carbonylation reaction is not limited to methanol as substrate. Thus, the successful carbonylation of a variety of primary, secondary and even tertiary alcohols has been reported with the rhodium—methyl iodide catalyst system [5]. None of these reactions have, to our knowledge, been commercialized but they offer several interesting possibilities.

Propionic acid, for example, could be prepared by ethanol carbonylation (Reaction 22). This represents an interesting alternative to propionic acid manufacture via ethylene hydroformylation or hydroxycarbonylation (Chapters 4 and 5), especially if ethanol became cheaply available from methanol homologation (see below).

$$CH_3CH_2OH + CO \longrightarrow CH_3CH_2CO_2H \tag{22}$$

Similarly, carbonylation of benzylic alcohols could constitute a useful route to phenylacetic acids (c.f. carbonylation of benzyl halides, discussed in Chapter 5).

$$ArCH_2OH + CO \longrightarrow ArCH_2CO_2H \tag{23}$$

Indeed, it is safe to say that although the carbonylation of methanol has been extensively studied the scope of the alcohol carbonylation reaction as a synthetic procedure in organic chemistry has been only cursorily examined.

7.2 Acetic Anhydride from Methyl Acetate Carbonylation

The carbonylation of carboxylic acid esters, in the absence of water, affords the corresponding acid anhydrides [13–15]. Methyl acetate, for example, is converted to acetic anhydride (Reaction 24). The corresponding ether can also be used as the substrate (Reaction 25), in which case the reaction presumably involves the intermediacy of the ester.

$$CH_3CO_2CH_3 + CO \longrightarrow (CH_3CO)_2O \tag{24}$$

$$CH_3OCH_3 + 2\ CO \longrightarrow (CH_3CO)_2O \tag{25}$$

Reactions 24 and 25 are effectively catalysed by rhodium [13, 14] or nickel

[15] compounds in conjunction with methyl iodide and an amine or phosphine as co-catalysts. Acetic acid is used as the solvent and typical conditions are 175°C/25 bar (Rh) and 200°C/60 bar (Ni).

A study of the kinetics and mechanism of the rhodium system has been recently reported [16]. The reaction rate is independent of methyl acetate concentration and of the CO pressure above 15 bar. The reaction is first-order with respect to rhodium, methyl iodide and the base (R_3P or R_3N) if its concentration is low. Chromium compounds, which are usually added to the catalyst system [13] do not influence the rate but markedly reduce the induction period. It was concluded [15] that the chromium compound, e.g. $Cr(CO)_6$, facilitates the generation of the active catalyst from the $RhCl_3$.

The reaction mechanism depicted in Scheme 7.2 was proposed as being consistent with the experimental data. The active catalyst is $RhL_2(CO)I$, formed by reduction of $RhCl_3$ with CO in the presence of iodide and the organic base, L. The first step in the catalytic cycle is oxidative addition of methyl iodide. The next step, migratory insertion of CO, is facilitated by the organic base, R_3P or R_3N. This explains the dependence of the rate on the concentration, steric bulk and nucleophilic strength of the base. This is then followed by reductive elimination of acetyl iodide.

$$RhL_2(CO)I + CH_3I \longrightarrow CH_3RhL_2(CO)I_2$$

$$\xrightarrow{L} CH_3CORhL_3I_2 \longrightarrow CH_3COI + RhL_3I$$

$$\xrightarrow[-L]{CO} RhL_2(CO)I$$

Scheme 7.2

Up to this point the mechanism is essentially the same as that of methanol carbonylation (see Scheme 7.1). However, in the absence of water and methanol the acetyl iodide reacts with acetate ion (acetic acid) to give acetic anhydride and complete the catalytic cycle:

$$CH_3COI + CH_3CO_2^- \longrightarrow (CH_3CO)_2O + I^- \tag{26}$$

$$CH_3CO_2CH_3 + I^- \longrightarrow CH_3I + CH_3CO_2^- \tag{27}$$

A similar mechanism is probably applicable to the nickel system.

According to recent reports [17] this carbonylation technology will be commercialized by Eastman Kodak in a new syn gas-based acetic anhydride plant due to go on stream in 1983. The new plant will produce 1 billion pounds of acetic anhydride per year. It will replace an existing plant which produces acetic anhydride via the high temperature pyrolysis of acetic acid to ketene (the acetic acid is derived from ethylene). The carbonylation route is much less energy intensive than the ketene route and is independent of oil-based hydro-carbon feedstocks.

The acetic anhydride will be used to make cellulose acetate. The acetic acid generated in this reaction is then reacted with methanol to give the methyl acetate needed for the carbonylation step. The overall reaction amounts to the conversion of cellulose, methanol and CO to cellulose acetate:

$$\text{cellulose} + Ac_2O \longrightarrow \text{cellulose acetate} + HOAc \tag{28}$$

$$HOAc + MeOH \longrightarrow MeOAc + H_2O \tag{29}$$

$$MeOAc + CO \longrightarrow Ac_2O \tag{30}$$

Net reaction:

$$\text{cellulose} + CO + MeOH \longrightarrow \text{cellulose acetate} + H_2O \tag{31}$$

The methanol needed for Reaction 29 will be made from syn gas in a methanol synthesis unit. The syn gas in turn will be generated by coal gasification. Some of the carbon monoxide from the gasifier is separated and used in the carbonylation step. Overall it constitutes a choice example of a completely integrated, coal-based chemicals complex (Scheme 7.3). An additional feature is that the sulphur recovered from the coal gasifier will be utilised in sulphuric acid manufacture. In the future such a coal-based chemicals complex could be further expanded to include other syn gas/methanol-based products.

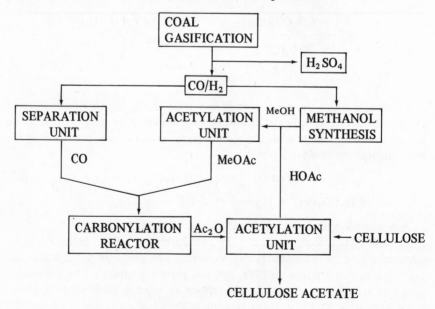

Scheme 7.3

7.3 Vinyl Acetate from Methyl Acetate

Halcon workers [18], following on from their success with acetic anhydride synthesis via methyl acetate carbonylation, reported a process for the manufacture of vinyl acetate by the following steps:

$$MeOAc + CO \xrightarrow[R_3N \text{ or } R_3P]{Rh/MeI} Ac_2O \qquad (32)$$

$$CH_3CHO + Ac_2O \longrightarrow CH_3CH(OAc)_2 \qquad (33)$$

$$CH_3CH(OAc)_2 \xrightarrow{\Delta} CH_2{=}CHOAc + HOAc \qquad (34)$$

$$MeOH + HOAc \longrightarrow MeOAc + H_2O \qquad (35)$$

Net reaction: $CH_3CHO + CO + MeOH$

$$\longrightarrow CH_2{=}CHOAc + H_2O \qquad (36)$$

The novel step in this process is methyl acetate carbonylation, since Reactions 33–35 were known technology. It was subsequently found [19] that when carbon monoxide is replaced with syn gas, reaction with methyl acetate in the presence of Rh or Pd, methyl iodide and an organic base, affords ethylidene diacetate directly according to the stoichiometry:

$$2\,MeOAc + 2\,CO + H_2 \xrightarrow[\substack{R_3N \text{ or } R_3P \\ 150°C/60\ bar}]{Rh \text{ or } Pd/MeI} CH_3CH(OAc)_2 + HOAc \qquad (37)$$

This direct route obviates the need for acetaldehyde as a raw material. Subsequent pyrolysis of the ethylidene acetate to vinyl acetate and acetic acid (Reaction 34) is established technology. Recycling of the acetic acid with methanol to form methyl acetate results in a net conversion of methanol and syn gas to the important industrial monomer, vinyl acetate.

$$2\,CH_3OH + 2\,CO + H_2 \longrightarrow CH_2{=}CHOAc + 2\,H_2O \qquad (38)$$

The mechanism of Reaction 37 has not been discussed but it probably involves initial carbonylation of the methyl acetate to acetic anhydride (Reaction 32) followed by catalytic hydrogenolysis of the latter to acetaldehyde and acetic acid. The acetaldehyde then reacts with unconverted acetic anhydride to give ethylidene diacetate (Reaction 33). The catalytic hydrogenolysis of acetic anhydride to acetaldehyde (Reaction 39) has been reported to occur with syn gas in the presence of $Co_2(CO)_8$ [20] or $Pd(OAc)_2/MeI/Bu_3P$ [21] as catalyst:

$$(CH_3CO)_2O + H_2 \longrightarrow CH_3CHO + CH_3CO_2H \qquad (39)$$

More recently, the high yield conversion of acetic anhydride to ethylidene acetate using a supported palladium catalyst and acetyl chloride promotor has been reported [22]. Ethylidene acetate is produced in 98% selectivity at 130°C and 35 bar. The acetyl chloride promotor can be added directly or generated *in situ* by the addition of hydrogen chloride. The reaction presumably involves the following steps:

$$CH_3COCl + Pd \longrightarrow CH_3COPdCl \qquad (40)$$

$$CH_3COPdCl + H_2 \longrightarrow CH_3CHO + Pd + HCl \qquad (41)$$

$$(CH_3CO)_2O + HCl \rightleftharpoons CH_3COCl + CH_3CO_2H \qquad (42)$$

$$CH_3CHO + (CH_3CO)_2O \longrightarrow CH_3CH(O_2CCH_3)_2 \qquad (43)$$

Net reaction:

$$2 (CH_3CO)_2O + H_2 \longrightarrow CH_3CH(O_2CCH_3)_2 + CH_3CO_2H \quad (44)$$

Similarly, in the $Co_2(CO)_8$-catalysed hydrogenolysis of acetic anhydride the initial step is probably the formation of an acylcobalt intermediate by reaction of acetic anhydride with the strong acid, $HCo(CO)_4$.

$$(CH_3CO)_2O + HCo(CO)_4 \longrightarrow CH_3COCo(CO)_4 + CH_3CO_2H \qquad (45)$$

This is followed by hydrogenolysis of the acylcobalt intermediate to acetaldehyde and $HCo(CO)_4$. The reaction is applicable to other anhydrides, e.g. cyclic anhydrides afford carboxyaldehydes [20]:

$$(CH_2)_n \underset{CO}{\overset{CO}{\diagdown}} O + H_2 \xrightarrow{[Co_2(CO)_8]} (CH_2)_n \underset{CHO}{\overset{CO_2H}{\diagup}} \qquad (46)$$

7.4 Propionic and Methacrylic Acids

As mentioned earlier, propionic acid can be prepared by carbonylation of ethanol (Reaction 22). Ethanol can in turn be prepared by homologation of methanol with syn gas (see next section). Propionic acid is itself not an important bulk chemical. However, if readily available it could become an important intermediate for methacrylic acid manufacture via condensation with formaldehyde (see preceding chapter). Since methanol and formaldehyde are also derived from syn gas, the overall process constitutes an all-syn gas route to methacrylic acid or methyl methacrylate as shown in Scheme 7.4.

Methyl methacrylate is currently manufactured from acetone following the cyanohydrin route:

$$(CH_3)_2CO + HCN \longrightarrow (CH_3)_2C(OH)CN \xrightarrow[H_2SO_4]{MeOH} CH_2{=}C(CH_3)CO_2Me \quad (47)$$

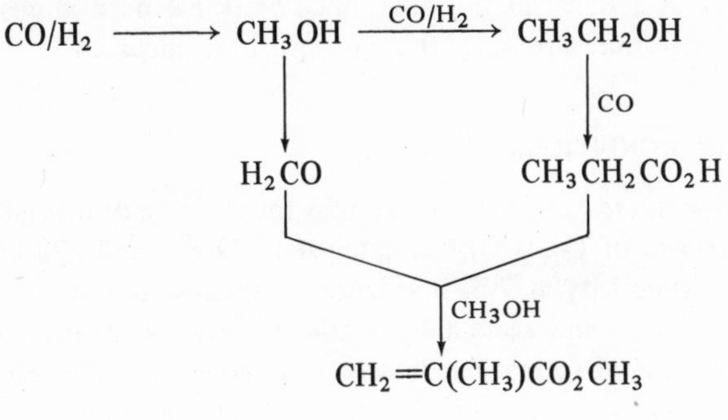

$$CO/H_2 \longrightarrow CH_3OH \xrightarrow{CO/H_2} CH_3CH_2OH$$

$$\downarrow \qquad\qquad\qquad \downarrow CO$$

$$H_2CO \qquad\qquad\qquad CH_3CH_2CO_2H$$

$$\downarrow CH_3OH$$

$$CH_2=C(CH_3)CO_2CH_3$$

Scheme 7.4

7.5 Methanol Homologation

When the cobalt carbonyl-catalysed carbonylation of methanol is carried out in the presence of hydrogen, i.e. using syn gas instead of pure CO, this can result in the formation of ethanol as a major product according to the stoichiometry:

$$CH_3OH + CO + 2\,H_2 \xrightarrow{[Co_2(CO)_8]} CH_3CH_2OH + H_2O \qquad (48)$$

This reaction was first reported in the 1940s [23–26] and the term homologation was introduced by Wender [25, 26]. This has found wider acceptance than the alternative name, hydrocarbonylation. The reaction was briefly studied at the time but then interest remained dormant for more than twenty years [27, 28]. The recent revival of interest is a result of the potential importance of Reaction 48 as a key step in alternative, coal-based ethylene manufacture (Scheme 7.5).

$$Coal \longrightarrow CO/H_2 \longrightarrow CH_3OH$$

$$\xrightarrow[-H_2O]{CO/H_2} CH_3CH_2OH \xrightarrow[-H_2O]{} CH_2=CH_2$$

Scheme 7.5

The subsequent step, ethanol dehydration, is well-established technology although it has attracted much attention recently [29,30], mainly in connection with ethylene from biomass projects. If ethanol becomes readily available from syn gas in the future it could also be directly converted to several bulk chemicals such as acetaldehyde and butadiene (see below).

The literature up to 1979 on methanol homologation was reviewed recently by Slocum [31] and Bahrmann and Cornils [32]. However, as a result of the

current flourish of activity in this area much work has been published, particularly in the patent literature, since these reviews have appeared.

CATALYSTS AND CONDITIONS

Wender and co-workers [25, 26] showed that the reaction of methanol with syn
gas, in the presence of $Co_2(CO)_8$ as catalyst at 185°C and 270 bar, afforded
ethanol in 39% selectivity at 70% methanol conversion. Important co-products
were acetic acid esters and acetaldehyde and the latter was presumed to be an
intermediate in ethanol formation. It was later shown that the reaction is promoted by iodide ion [33–35]. Selectivity improvements were achieved by the
addition of ruthenium compounds [36–39] and/or phosphine ligands [37–42].

In the current drive to develop a commercially viable process for methanol
homologation three aspects are of importance when considering the various
catalyst systems: activity, selectivity and stability. In other words the goal is to
maximize the selectivity and catalyst stability at an acceptable conversion rate.
Iodide ion, for example, increases the rate of reaction but not the selectivity to
ethanol.

Ruthenium compounds, on the other hand, increase the selectivity to
ethanol. This is a result of their higher hydrogenation activity, compared to
cobalt, which allows for facile reduction of the acetaldehyde intermediate.
Ruthenium compounds, in conjunction with iodide promotors, are also catalysts
in the absence of cobalt [43].

The beneficial effect of phosphine ligands on selectivity is also attributable to
their enhancing effect on the hydrogenation activity of cobalt carbonyl catalysts
(compare the effect of phosphine ligands on cobalt hydroformylation catalysts,
discussed in Chapter 4). Phosphine ligands also have a beneficial effect on the
catalyst stability [39] and prevent its decomposition to cobalt metal. Excellent
selectivities (*ca.* 90%) have been recently reported for a cobalt iodide/bis-
diphenylphosphinoalkane ligand system [42].

The best results appear to accrue from the use of $Co/I^-/R_3P$ or $Co/Ru/I^-/R_3P$
catalyst systems at *ca.* 200°C and 200 bar (see Table 7.I for a comparison of the
various systems). Significant improvements have also been claimed with the
original $Co_2(CO)_8$ system when the process is carried out in continuous operation [44]. Similarly, improvements have also been claimed to accrue from the
use of inert solvents, particularly dioxane, with both the $Co_2(CO)_8$ [45] and
CoI_2 [46] systems. However, caution should be exercised in interpreting results
obtained in such ether solvents that, in principle, can be cleaved to ethanol under
the reaction conditions [47].

In contrast to what is observed in olefin hydroformylations and alcohol
carbonylations, rhodium compounds are poor catalysts for alcohol homologa-

TABLE 7.I

Methanol homologation. A comparison of various catalysts

Catalyst	Solvent	Pressure (bar)	Temp. (°C)	MeOH conc. (%)	Molar selectivity (%)							Reference
					EtOH	EtOH + precursors[a]	MeOAc	EtOAc	Et$_2$O	CH$_4$	Others	
CoI$_2$/Bu$_3$P	n-octane	150	200	31	62	87	7	–	–	2	4	40
Co(acac)$_2$[b]	none	300	190	31	52	64	8	–	0.2	6	ca. 22	44
Co(acac)$_2$/I$_2$/Ph$_3$P	none	270	175	48	58	58	18	–	13	–	11	37
Ru(acac)$_3$/Ph$_3$P												
CoI$_2$/ Ph$_2$P(CH$_2$)$_6$PPh$_2$	benzene/ water	180	200	28	85	90	0.3	–	–	5	ca. 5	42
Co$_2$(CO)$_8$	dioxane	310	180	53	69	81	2	–	–	–	17	45
Co(OAc)$_2$/I$_2$	dioxane	200	190	93	69	75	20	5	–	–	–	46
Co(OAc)$_2$/I$_2$/ [Ph$_3$PCo(CO)$_3$]$_2$/	none	200	180	43	80	80	–	–	–	–	20	39
RuCl$_3$/C$_6$H$_{11}$)$_3$P I$_2$/Ru(acac)$_3$	none	270	200	44	72	72	10	–	3	–	15	38

a EtOH precursors are acetaldehyde and acetaldehyde dimethyl acetal.
b In continuous operation.

logation, even in the presence of iodide promotors. This can be attributed to the low hydrogenation activity of rhodium catalysts. Thus, we have already noted (see Section 7.1) that rhodium/iodide catalysts effect the carbonylation of methanol to acetic acid even in the presence of relatively large amounts of hydrogen. Homologation is observed with rhodium catalysts only at very high (40 : 1) H_2 : CO molar ratios [48].

The present state of the art is summarized in Table 7.I. Direct comparison of the various catalyst systems is difficult as the selectivities are often reported at different conversion levels. When comparing selectivities it is also important to bear in mind that certain ethanol precursors (acetaldehyde and its acetals) can, in principle, be recycled. Generally, selectivities of 75–90% have been observed at conversion levels of 30–90%. According to recent reports [49], Gulf workers have fine-tuned the $Co/Ru/I_2/R_3P$ system to such an extent that ethanol selectivities above 90% are obtained.

MECHANISM

The following mechanism (Scheme 7.6) was proposed by Wender [28] to account for the $Co_2(CO)_8$-catalysed homologation of methanol in the absence of iodide ion:

$$Co_2(CO)_8 + H_2 \longrightarrow HCo(CO)_4 \qquad\qquad (49)$$

$$HCo(CO)_4 + CH_3OH \longrightarrow CH_3Co(CO)_4 + H_2O \qquad\qquad (50)$$

$$CH_3Co(CO)_4 + CO \longrightarrow CH_3COCo(CO)_4 \qquad\qquad (51)$$

$$CH_3COCo(CO)_4 + H_2 \longrightarrow CH_3CHO + HCo(CO)_4 \qquad\qquad (52)$$

Or:

$$CH_3COCo(CO)_4 + HCo(CO)_4 \longrightarrow CH_3CHO + Co_2(CO)_8 \quad (53)$$

$$CH_3CHO + H_2 \xrightarrow{[HCo(CO)_4]} CH_3CH_2OH \qquad\qquad (54)$$

The active catalyst is $HCo(CO)_4$, formed via reduction of $Co_2(CO)_8$ with hydrogen. The rate-limiting step in the process is the formation of the methyl-cobalt intermediate by reaction of $HCo(CO)_4$ with methanol (Reaction 50). This reaction may be compared with, for example, the formation of methyl bromide from methanol and HBr, and proceeds through the S_N2 displacement of water from protonated methanol:

$$CH_3OH + HCo(CO)_4 \rightleftharpoons CH_3\overset{+}{O}H_2 + Co(CO)_4^-$$

$$\longrightarrow CH_3Co(CO)_4 + H_2O \qquad\qquad (55)$$

Reaction 55 is possible because $HCo(CO)_4$ is a very strong acid, capable of protonating methanol. This explains the superior activity of cobalt compared to other Group VIII metals as a catalyst for the alcohol homologation reaction. Other metals, such as Rh and Ru, form only weakly acidic carbonyl hydrides. Indeed, in the absence of iodide ion $Ru_3(CO)_{12}$ catalyses the selective formation of methyl formate from methanol and carbon monoxide, and no homologation is observed [43].

Reaction 55 is followed by CO insertion into the methylcobalt intermediate to give an acetylcobalt species. Hydrogenolysis of the latter with H_2 or $HCo(CO)_4$ gives acetaldehyde [50]. The final step is the reduction of acetaldehyde to ethanol. The *stoichiometric* reduction of aldehydes with $HCo(CO)_4$ was shown to proceed readily at ambient temperature in hexane as solvent [51].

$$RCHO + 2\,HCo(CO)_4 \longrightarrow RCH_2OH + Co_2(CO)_8 \qquad (56)$$

Considering the polarization of the two reactants it would seem likely that Reaction 56 involves attack of the tetracarbonylcobaltate anion at the positively charged carbon of the aldehyde group as shown in Scheme 7.7.

$$CH_3CHO + HCo(CO)_4 \rightleftharpoons CH_3CH(OH)^+ + Co(CO)_4^-$$

$$\longrightarrow \underset{\overset{|}{CH_3CH-Co(CO)_4}}{\overset{OH}{\;}} \xrightarrow{HCo(CO)_4} CH_3CH_2OH + Co_2(CO)_8$$

Scheme 7.7

$HCo(CO)_4$ is not an exceptional hydrogenation catalyst and improvements in selectivity are, therefore, obtained by the addition of superior hydrogenation catalysts, such as ruthenium compounds as noted above. The addition of heterogeneous hydrogenation catalysts, particularly those based on rhenium, has also been claimed to result in improved selectivity [52]. In contrast, the reaction can be stopped at the acetaldehyde stage by a suitable choice of ligand. Thus, reaction of methanol with syn gas in the presence of $[Ar_3AsCo(CO)_4]_2$ and iodine results in the formation of acetaldehyde as the major (50–55%) product, together with smaller amounts (*ca.* 20%) of ethanol [41, 53].

In the presence of *iodide as co-catalyst* methanol is converted to methyl iodide via:

$$CH_3OH + HI \longrightarrow CH_3I + H_2O \qquad (57)$$

The rate-limiting step is then displacement of iodide by the cobalt carbonyl anion (Reaction 58). The promoting effect of iodide is thus due to the fact that Reaction 58 is more favourable than Reaction 50.

$$CH_3I + HCo(CO)_4 \longrightarrow CH_3Co(CO)_4 + HI \qquad (58)$$

The major by-products observed in methanol homologation are generally acetate esters, acetals, ethers and methane. Their modes of formation are summarized schematically below (Scheme 7.8).

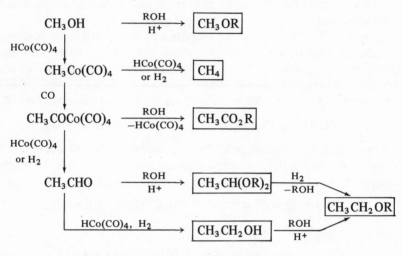

Scheme 7.8

Methyl and ethyl acetate result from competing alcoholysis of the acetyl-cobalt intermediate. The selectivity to ethanol is, thus, influenced by the relative rates of hydrogenolysis and alcoholysis of the intermediate. It should be noted, however, that if the hydrogenation activity of the catalyst system is too high this will favour the formation of methane by hydrogenolysis of the methylcobalt intermediate.

Acetals are formed from the acid-catalysed reaction of methanol or ethanol with the acetaldehyde intermediate in competition with its further hydrogenation. Ethers can result from the acid-catalysed dehydration of the corresponding alcohols and/or the hydrogenolysis of acetals (see Scheme 7.8).

In addition to these major by-products small amounts of formate esters and higher alcohols are often observed. The latter result from slow homologation of the ethanol product.

7.6 Homologation of Other Alcohols

Berty and co-workers [33] studied the relative rates of homologation of various alcohols in the presence of cobalt carbonyl and iodine. Consistent with rate-limiting S_N2 displacement of iodide by $Co(CO)_4^-$, ethanol reacts a factor 50 times more slowly than methanol. This explains why methanol is selectively converted to ethanol with negligible accompanying formation of n-propanol by further reaction. Isopropanol is, as expected in an S_N2 mechanism, similarly unreactive. Tertiary butanol, in contrast, exhibits high reactivity, suggesting a change in mechanism to S_N1 formation of the tertiary butyl cation via:

$$(CH_3)_3COH + HCo(CO)_4 \longrightarrow (CH_3)_3C^+ + H_2O + Co(CO)_4^- \quad (59)$$

Benzyl alcohol exhibits roughly the same reactivity as methanol [33]. The rate of $HCo(CO)_4$-catalysed homologation of a series of benzylic alcohols is markedly increased by electron-releasing substitutuents in the benzene ring, consistent with nucleophilic displacement being the rate-limiting step [54]. For example, p-methoxybenzyl alcohol is 10^4 times as reactive as benzyl alcohol [54]. The yields of the expected homologation product, 2-arylethanol, are only poor to moderate (15–45%) and substantial amounts of the corresponding toluene are formed as shown below:

$$ArCH_2OH \underset{\xrightarrow{\quad H_2 \quad} ArCH_3 + H_2O \quad (61)}{\overset{\xrightarrow{\quad H_2/CO \quad} ArCH_2CH_2OH + H_2O \quad (60)}{\Big\langle}}$$

CONVERSION OF TOLUENE TO STYRENE

The homologation of benzyl alcohol to 2-phenylethanol is of potential commercial value as the key step in a proposed toluene/syn gas route to styrene [55] via the following sequence:

$$PhCH_3 \xrightarrow{\quad O_2 \quad} PhCH_2OH \; (+ PhCHO) \quad (62)$$

$$PhCH_2OH \xrightarrow{\quad CO/H_2 \quad} PhCH_2CH_2OH + H_2O \quad (63)$$

$$PhCH_2CH_2OH \xrightarrow{\quad -H_2O \quad} PhCH{=}CH_2 + H_2O \quad (64)$$

The autoxidation of toluene affords a mixture of mainly benzyl alcohol and benzaldehyde. The formation of the latter does not have serious consequences as it can be hydrogenated to benzyl alcohol under the homologation conditions. As noted above, the $HCo(CO)_4$-catalysed homologation of benzyl alcohol affords 2-phenylethanol in 31% yield together with 63% toluene. However, it was recently shown that the selectivity is markedly improved using a cobalt/ruthenium/iodide catalyst combination with small amounts of water added. Using this system at 270 bar and 130°C, 2-phenylethanol was formed in 79% selectivity at 47% benzyl alcohol conversion [56].

In a modified version of this process [57] toluene is subjected to palladium-catalysed oxidation in acetic acid to give a mixture of benzyl acetate and benzaldehyde. This mixture is partially hydrolysed and the resulting mixture of benzyl alcohol, benzyl acetate and benzaldehyde reacted with syn gas to give mainly 2-phenylethanol and 2-phenylethyl acetate in high selectivity. The subsequent conversion of this mixture to styrene, via elimination of water and acetic acid, involves established technology.

$$PhCH_3 \xrightarrow[\text{2. } H_2O]{\text{1. } O_2, \text{HOAc, } [Pd^{II}]} PhCH_2OAc + PhCH_2OH + PhCHO$$

$$\xrightarrow[\text{[Co/Ru/I}^-]]{\text{CO/H}_2} PhCH_2CH_2OH + PhCH_2CH_2OAc \xrightarrow{-H_2O, \text{ HOAc}} PhCH=CH_2$$

Scheme 7.9

This constitutes a potentially attractive route to styrene. The process that currently accounts for the manufacture of styrene (Reaction 65) involves the use of benzene and ethylene, which are more expensive raw materials than toluene and syn gas.

$$PhH + CH_2=CH_2 \xrightarrow{H^+} PhCH_2CH_3 \xrightarrow{-H_2} PhCH=CH_2 \qquad (65)$$

7.7 Homologation of Ethers and Esters

In the course of studying the homologation of methanol in the presence of ruthenium catalysts and iodide promotors Braca and co-workers [43, 58, 59] made an interesting observation. They found that dimethyl ether and methyl acetate, which, together with ethanol, were the primary products of the reaction, could be further carbonylated. It was subsequently verified in independent experiments that dimethyl ether and methyl acetate are converted to ethyl acetate and the following reaction stoichiometries were proposed.

$$CH_3OCH_3 + 2\,CO + 2\,H_2 \longrightarrow CH_3CO_2C_2H_5 + H_2O \qquad (66)$$

$$CH_3CO_2CH_3 + CO + 2\,H_2 \longrightarrow CH_3CO_2C_2H_5 + H_2O \qquad (67)$$

The best results were obtained in acetic acid or, better, acetic anhydride as solvent at 200°C and 150 bar. Selectivities to ethyl acetate were as high as 80%. Ruthenium compounds appear to be unique catalysts for Reaction 66 and 67 as these reactions are not observed with Co and Rh catalysts. When dimethyl ether is the substrate the initial step is carbonylation to methyl acetate.

$$CH_3OCH_3 + CO \xrightarrow{[Ru/I^-]} CH_3CO_2CH_3 \qquad (68)$$

Since Reaction 68 is also catalysed by Rh and Ni compounds (see Section 7.2) the uniqueness of Ru apparently lies in the subsequent conversion of methyl acetate via Reaction 67. The mechanism of this reaction has not yet been clearly defined. By analogy with the rhodium-based systems described earlier a possible reaction pathway could involve the steps shown in Scheme 7.10.

$$CH_3CO_2CH_3 \xrightarrow{CO} (CH_3CO)_2O \xrightarrow{H_2} CH_3CHO + CH_3CO_2H$$

$$CH_3CHO \xrightarrow{H_2} CH_3CH_2OH \xrightarrow{(CH_3CO)_2O} CH_3CO_2C_2H_5 + CH_3CO_2H$$

Scheme 7.10

The active catalyst is presumably a ruthenium carbonyl species and the various steps in Scheme 7.10 are assumed to involve methyl and acetyl-ruthenium compounds as transitory intermediates. The essential difference between the ruthenium system and the rhodium one described earlier (see Section 7.2) appears to lie in the fate of the acetaldehyde intermediate. With rhodium-based catalysts the acetaldehyde is scavenged by the acetic anhydride to give ethylidene diacetate as the major product. Ruthenium catalysts, on the other hand, are superior hydrogenation catalysts and promote the smooth reduction of acetaldehyde to ethanol. Subsequent reaction of the latter with acetic anhydride (or acetic acid) leads to the formation of ethyl acetate as the major product.

If ethyl acetate is formed by reaction of ethanol with acetic anhydride, as implied in Scheme 7.10, this leads to the overall stoichiometry shown in Equation 69 rather than that shown in Equation 67.

$$2\,CH_3CO_2CH_3 + 2\,CO + 2\,H_2 \longrightarrow CH_3CO_2C_2H_5 + 2\,CH_3CO_2H \quad (69)$$

It is worth noting in this context that the highest selectivities to ethyl acetate were observed in the presence of added acetic anhydride as would be expected for the reaction pathway shown in Scheme 7.10. Moreover, the homologation of methyl acetate in toluene as solvent [59] led to the formation of ethyl acetate and acetic acid in proportions consistent with the stoichiometry of Reaction 69. In practice some of the acetic acid formed may also be hydrogenated to ethyl derivatives (ethanol, ethyl acetate) which leads to an overall stoichiometry intermediate between that of Reactions 67 and 69.

Recently mixed metal catalyst systems, consisting of $Co_2(CO)_8/Ru_3(CO)_{12}/CH_3I$ [60] and $RhCl_3/RuCl_3/ZnI_2$ [61], have also been reported to promote the homologation of methyl acetate. Mixtures of ethyl acetate and acetic acid were formed roughly according to the stoichiometry of Equation 69. With the second catalyst system improved yields were observed in the presence of a crown ether such as 18-crown-6.

When the acetic acid formed in Reaction 69 is recycled by reacting it with methanol the overall result amounts to the conversion of methanol and syn gas to ethyl acetate as shown below:

$$2\,CH_3OH + 2\,CH_3CO_2H \longrightarrow 2\,CH_3CO_2CH_3 + 2\,H_2O \tag{70}$$

$$2\,CH_3CO_2CH_3 + 2\,CO + 2\,H_2 \longrightarrow CH_3CO_2C_2H_5 + 2\,CH_3CO_2H \tag{71}$$

Net reaction:

$$2\,CH_3OH + 2\,CO + 2\,H_2 \longrightarrow CH_3CO_2C_2H_5 + 2\,H_2O \tag{72}$$

This provides a potentially attractive route to the industrial solvent ethyl acetate that is currently manufactured from ethylene-based raw materials. Subsequent pyrolysis of the ethyl acetate to ethylene provides a syn gas-based route to the latter that is a possible alternative to the methanol homologation route described earlier. This route is illustrated in Scheme 7.11.

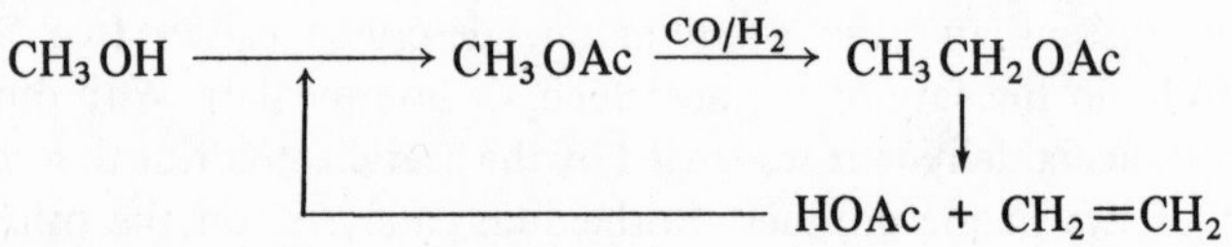

$$CH_3OH \longrightarrow CH_3OAc \xrightarrow{CO/H_2} CH_3CH_2OAc$$
$$\downarrow$$
$$HOAc + CH_2{=}CH_2$$

Scheme 7.11

Another interesting variation on this theme is the recently reported two-step process for the preparation of ethylene by reaction of syn gas with a carboxylic acid, to form the corresponding ethyl ester, followed by pyrolysis of the latter [62].

$$RCO_2H + 2\,CO + 4\,H_2 \xrightarrow{[Ru/I^-]} RCO_2Et + 2\,H_2O \tag{73}$$

$$RCO_2Et \longrightarrow RCO_2H + CH_2{=}CH_2 \tag{74}$$

Reaction 73 is catalysed by ruthenium compounds in the presence of halide co-catalysts at 220°C and 400 bar. The best results are obtained with tetraalkyl-phosphonium bromides as co-catalysts. For example, with RuO_2 and heptyl-triphenylphosphonium bromide as the catalyst combination, propionic acid is converted to a mixture of mainly ethyl propionate (38 wt%) and smaller amounts of methyl propionate (17 wt%) and n-propyl propionate (8 wt%).

The mechanism of the reaction was not discussed. One possible pathway involves the initial formation of methanol from syn gas, which is known to be catalysed by ruthenium compounds. This is then followed by a series of oxidative addition, CO insertion and hydrogenation steps as outlined in Scheme 7.12.

We note that the formation of ethyl esters by acidolysis of the putative ethyl

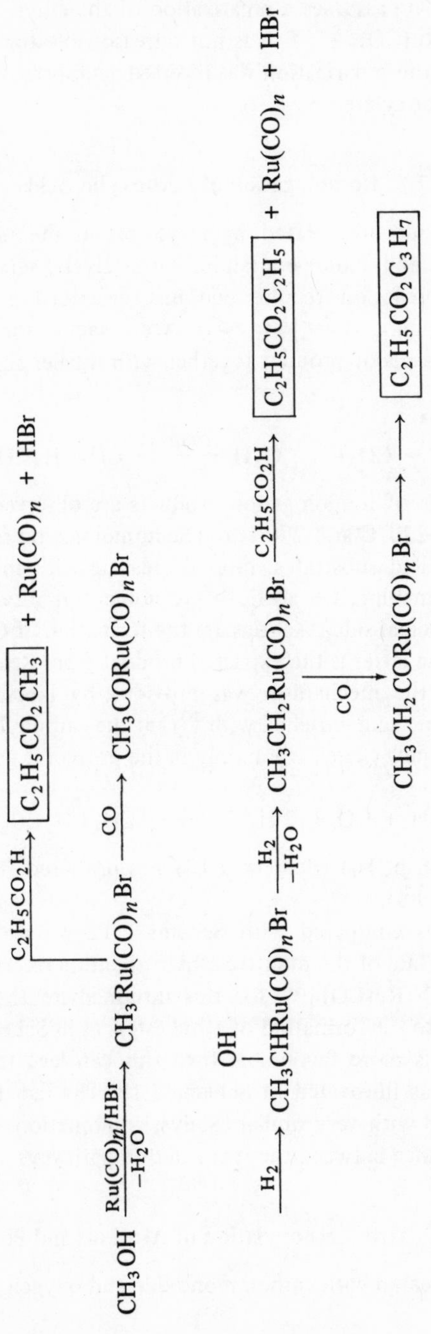

$$CH_3OH \xrightarrow[-H_2O]{Ru(CO)_n/HBr} CH_3Ru(CO)_nBr \xrightarrow{CO} CH_3CORu(CO)_nBr \xrightarrow{C_2H_5CO_2H} \boxed{C_2H_5CO_2CH_3} + Ru(CO)_n + HBr$$

$$\xrightarrow{H_2} CH_3\overset{\displaystyle OH}{\overset{|}{C}}HRu(CO)_nBr \xrightarrow[-H_2O]{H_2} CH_3CH_2Ru(CO)_nBr \xrightarrow{CO} CH_3CH_2CORu(CO)_nBr \xrightarrow{C_2H_5CO_2H} \boxed{C_2H_5CO_2C_2H_5} + Ru(CO)_n + HBr$$

$$CH_3CH_2CORu(CO)_nBr \longrightarrow \boxed{C_2H_5CO_2C_3H_7}$$

Scheme 7.12

ruthenium intermediate assumes a polarization of the ethyl–ruthenium bond in the direction: $Et^{\delta+}Ru(CO)_n{}^{\delta-}$. This is not unreasonable for an alkylruthenium carbonyl species. If the polarization was reversed acidolysis would afford ethane and a ruthenium carboxylate.

7.8 Homologation of Carboxylic Acids

When carboxylic acids are treated with syn gas in the presence of soluble ruthenium catalysts and iodide-containing co-catalysts, such as HI or methyl iodide, a different reaction to the one just described is observed, namely carboxylic acid homologation [63, 64]. Acetic acid, for example, affords propionic acid as the major product together with smaller amounts of n-butyric and n-valeric acids:

$$CH_3CO_2H \xrightarrow{\;CO/H_2\;} CH_3CH_2CO_2H \xrightarrow{\;CO/H_2\;} CH_3CH_2CH_2CO_2H, \text{etc} \quad (75)$$

The highest yields of homologation products are observed using the $RuO_2/$ MeI combination at 220°C and 270 bar. The homologation reaction is a general one for carboxylic acid substrates, rates decreasing with increasing molecular weight and with branching, i.e. acetic > propionic > n-butyric > isobutyric > pivalic acid. The principal side reactions are the formation of CO_2, hydrocarbons and ethyl acetate. The latter is the expected product from Reaction 73.

An insight into the mechanism was provided by labelling studies. Thus, homologation of acetic acid enriched with ^{13}C at the carbonyl carbon led to ^{13}C enrichment at the α-methylene carbon only in the propionic acid product, i.e.,

$$CH_3\overset{*}{C}O_2H + CO + 2H_2 \longrightarrow CH_3\overset{*}{C}H_2CO_2H + H_2O \quad (76)$$

The mechanism on p. 161 (Scheme 7.13) was proposed [63] to account for the experimental results.

If Scheme 7.13 is compared with Scheme 7.12 we see that the essential difference lies in the fate of the putative ethylruthenium intermediate. Assuming polarization of $Et^{\delta+} Ru(CO)_n{}^{\delta-}$ for this intermediate then acidolysis with RCO_2H should lead to the formation of ethyl esters as in Scheme 7.12. If carbon monoxide insertion is more favourable then this can lead to homologation of the carboxylic acid as illustrated in Scheme 7.13. The fact that both types of reaction are observed with very similar catalyst combinations suggests that there is a very delicate balance between the two reaction pathways.

7.9 Oxidative Carbonylation of Alcohols and Phenols

When alcohols are treated with carbon monoxide and oxygen, in the presence of

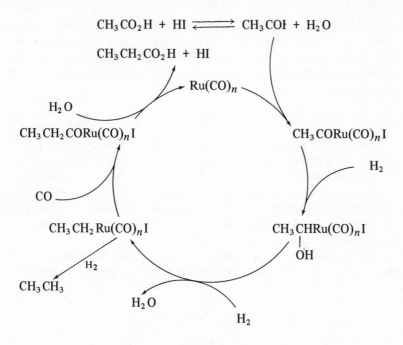

$$CH_3CO_2H + HI \rightleftharpoons CH_3COI + H_2O$$

Scheme 7.13

certain catalysts, *oxidative carbonylation* is observed. Depending on the catalyst system used this can lead to the formation of dialkyl carbonates (Reaction 77) or dialkyl oxalates (Reaction 78).

$$2\ ROH + CO + \tfrac{1}{2}O_2 \longrightarrow (RO)_2CO + H_2O \qquad (77)$$

$$2\ ROH + 2\ CO + \tfrac{1}{2}O_2 \longrightarrow (CO_2R)_2 + H_2O \qquad (78)$$

Reaction 77 is catalysed by copper compounds [65]. The cuprous chloride-catalysed oxidative carbonylation of methanol, for example, was used for the synthesis of dimethyl carbonate [65]. The latter has potential value as a substitute for the toxic reagents phosgene and dimethyl sulphate in a variety of applications [65]. Dimethyl carbonate formation occurs in two stages. In the first stage a methoxycupric species, MeOCuCl, is formed via:

$$2\ CuCl + 2\ CH_3OH + \tfrac{1}{2}O_2 \longrightarrow 2\ CH_3OCuCl + H_2O \quad (79)$$

This is followed by insertion of CO to form a transitory methoxycarbonyl-cupric species which reacts with another molecule of MeOCuCl as shown:

$$CH_3OCuCl \xrightarrow{\ CO\ } CH_3OCOCuCl \xrightarrow{CH_3OCuCl} (CH_3O)_2CO + CuCl \qquad (80)$$

Reaction 78 is catalysed by palladium compounds in the presence of copper salts [66] or nitrogen oxides [67] as co-catalysts. A key feature of these processes is the insertion of CO into a palladium(II) alkoxide to give a *bis*-alkoxycarbonylpalladium(II) species, which then reductively eliminates the dialkyl oxalate product as shown:

$$Pd^{II}(OR)_2 + 2\,CO \longrightarrow Pd^{II}(CO_2R)_2 \longrightarrow Pd^0 + (CO_2R)_2 \qquad (81)$$

In support of this proposal, *bis*-alkoxycarbonylpalladium(II) complexes have recently been isolated and characterised [68]. For example, $Pd(CO_2CH_3)_2(Ph_3P)_2$ is stable for several hours in methanol at ambient temperature under CO pressure, but decomposes to dimethyl oxalate on warming to $50°C$. The formation of the alkoxycarbonyl intermediate can be envisaged as a migratory insertion of an alkoxide ligand at a palladium(II) carbonyl complex (c.f. migratory insertion of an alkyl ligand to form an acylmetal complex).

The function of the co-catalyst is to facilitate the re-oxidation of the palladium catalyst. With nitrogen oxides this proceeds via the following steps:

$$2\,NO + \tfrac{1}{2}\,O_2 \longrightarrow N_2O_3 \qquad (82)$$

$$N_2O_3 + 2\,ROH \longrightarrow 2\,RONO + H_2O \qquad (83)$$

$$Pd^0 + 2\,RONO \longrightarrow Pd(OR)_2 + 2\,NO \qquad (84)$$

Both liquid- and vapour-phase processes have been described [67] and typical conditions are $70°C$ and 70 bar. Potential commercial interest in the reaction stems from the possibility that the dialkyl oxalate product can be hydrogenated to ethylene glycol. This affords a two step process for the conversion of syn gas to ethylene glycol:

$$2\,ROH + 2\,CO + \tfrac{1}{2}\,O_2 \longrightarrow (CO_2R)_2 + H_2O \qquad (85)$$

$$(CO_2R)_2 + 3\,H_2 \longrightarrow HOCH_2CH_2OH + 2\,ROH \qquad (86)$$

Net reaction:

$$2\,CO + 3\,H_2 + \tfrac{1}{2}\,O_2 \longrightarrow HOCH_2CH_2OH + H_2O \qquad (87)$$

This constitutes yet another syn gas-based route to ethylene glycol to be compared with those already mentioned in Chapter 6 and the direct route to be discussed in Chapter 9.

Phenols also undergo oxidative carbonylation in the presence of palladium catalysts, an oxidation co-catalyst and a base, to give the corresponding diaryl carbonate [69–71].

$$2\,ArOH + CO + \tfrac{1}{2}\,O_2 \xrightarrow[\text{co-catalyst}]{Pd^{II},\,\text{base}} (ArO)_2CO + H_2O \qquad (88)$$

Initially tertiary amines were used as the base [69, 70]. More recently, superior results were obtained using sodium hydroxide in combination with a phase transfer catalyst [71]. For example, good yields were obtained using $PdBr_2/Mn(acac)_2/Bu_4NBr$ together with 50% aqueous NaOH and molecular sieves (to remove water) in methylene chloride at ambient temperature and atmospheric pressure [71]. The diaryl carbonate products are used in the manufacture of polycarbonates.

7.10 Summary

Syn gas-derived methanol promises to be a key building block for a broad spectrum of industrial chemicals. By an appropriate choice of catalyst and conditions the reactions of methanol and its simple derivatives with syn gas can be manipulated to afford a variety of oxygen-containing products. These are summarized in Scheme 7.14. Although the yields and conditions of some of these reactions still leave a lot to be desired it is worth remembering that many of the reactions were unknown a few years ago. Improvements are, therefore, likely to be forthcoming in the near future.

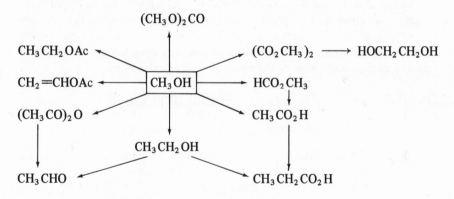

Scheme 7.14

Up until recently the catalysts of choice for homogeneous reactions of syn gas were generally cobalt or rhodium and, to a lesser extent, palladium and nickel. It is significant, therefore, that many of the new reactions described in this chapter employ ruthenium-based catalysts. The combination of good carbonylation and hydrogenation activity appears to make ruthenium compounds (in the presence of iodide promotors) particularly good catalysts for homologation reactions. We expect that this emphasis on ruthenium-based catalysts will continue in the future.

As an alternative to the direct conversion of methanol to oxygenates it can also be converted, via ethanol, to ethylene. Conversion of the latter to industrial derivatives can then be carried out using established technology. Significant improvements have been booked in the last few years with regard to the selectivity of methanol homologation to ethanol. In addition, other processes have recently been described that, in principle, could be used for the conversion of methanol to ethylene, e.g. via homologation of methyl acetate (see Scheme 7.15).

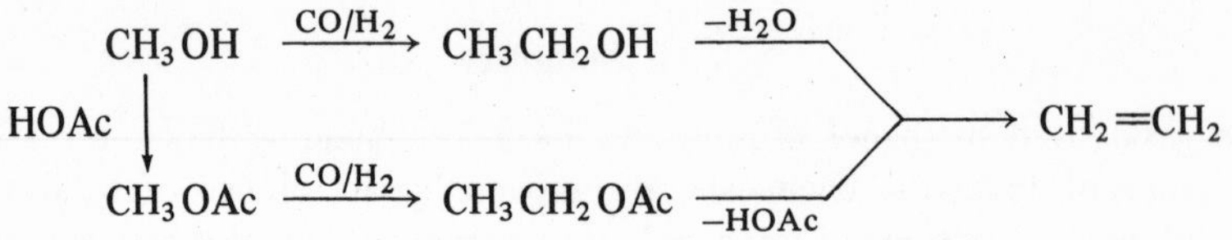

$$CH_3OH \xrightarrow{CO/H_2} CH_3CH_2OH \xrightarrow{-H_2O}$$

$$HOAc \downarrow$$

$$CH_3OAc \xrightarrow{CO/H_2} CH_3CH_2OAc \xrightarrow{-HOAc}$$

$$\longrightarrow CH_2{=}CH_2$$

Scheme 7.15

In the event that methanol homologation becomes an important source of ethylene a need may develop for alternative sources of other hydrocarbon feedstocks currently derived from naphtha cracking. In this context it is worth mentioning that during the Second World War butadiene was produced from fermentation ethanol. The process involves dehydrogenation to acetaldehyde followed by aldol condensation as shown in Scheme 7.16. If ethanol, derived from syn gas or biomass, becomes an important base chemical in the future this may lead to the resurrection of the ethanol-to-butadiene process.

$$CH_3CH_2OH \xrightarrow{-H_2} CH_3CHO \longrightarrow CH_3CH(OH)CH_2CHO$$

$$\xrightarrow{H_2} CH_3CH(OH)CH_2CH_2OH \xrightarrow{-2 H_2O} CH_2{=}CH{-}CH{=}CH_2$$

Net reaction: $2 CH_3CH_2OH \longrightarrow CH_2{=}CH{-}CH{=}CH_2 + 2 H_2O + H_2$

Scheme 7.16

References

1. R. P. Lowry and A. Aguilo, *Hydrocarbon Process.*, **53** (11), 103 (1974).
2. W. Reppe, *Justus Liebig's Ann. Chem.*, **582**, 1 (1953).
3. H. Hohenschutz, N. von Kutepow and W. Himmele, *Hydrocarbon Process.*, **45** (11), 141 (1966).
4. F. E. Paulik and J. F. Roth, *J. Chem. Soc. Chem. Commun.*, 1578 (1968).
5. F. E. Paulik, A. Hershman, W. R. Knox and J. F. Roth, *US Patent* 3,769,329 (1973) to Monsanto.
6. J. F. Roth, J. H. Craddock, A. Hershman and F. E. Paulik, *Chem. Tech.*, **1**, 600 (1971).
7. H. D. Grove, *Hydrocarbon Process.*, **51** (11), 76 (1972).

8. D. Forster, *J. Am. Chem. Soc.*, **98**, 846 (1976).
9. D. Forster, *Advan. Organometal. Chem.*, **17**, 255 (1979).
10. D. Forster, *J. Am. Chem. Soc.*, **97**, 951 (1975).
11. A. N. Haglieri and N. Rizkalla, *German Patent* 2,749,954 and 2,749,955 (1976) to Halcon International.
12. E. M. Thosteinson, B. D. Dombek and A. R. Fiato, *German Patent* 3,043,112 (1981) to Union Carbide.
13. N. Rizkalla, *German Patent* 2,610,036 (1976) to Halcon International.
14. H. Kuckertz, *German Patent* 2,450,965 (1976) to Hoechst.
15. I. Tomiya and Y. Kijima, *British Patent* 2,007,666 (1979) to Mitsubishi Gas.
16. M. Schrod and G. Luft, *Ind. Eng. Chem. Prod. Res. Dev.*, **20**, 649 (1981).
17. *Chemical Week*, January 16, 1980, p. 40.
18. C. G. Wan, *German Patent* 2,856,791 (1979) to Halcon International.
19. N. Rizkalla and C. N. Winnick, *German Patent* 2,610,035 (1976) to Halcon International.
20. K. N. Yamagami, H. Wakamatsu and J. Furakawa, *German Patent* 2,016,061 (1970) to Ajinomoto.
21. R. V. Porcelli, *German Patent* 2,952,517 (1980) to Halcon Research and Development.
22. J. W. Brockington and C. M. Bartish, *ACS Div. Petrol. Chem. Preprints*, **26** (3), 750 (1981).
23. G. Wietzel, K. Eder and A. Scheurmann, *German Patent* 867,849 (1953) to BASF; *C.A.* **48**, 1407i (1954).
24. R. E. Brooks, *US Patent* 2,457,204 (1948) to Du Pont.
25. I. Wender, R. Levine and M. Orchin, *J. Am. Chem. Soc.*, **71**, 4160 (1949).
26. I. Wender, R. A. Friedel and M. Orchin, *Science*, **113**, 206 (1951).
27. G. Albanesi, *Chim. Ind.* (Milan), **55**, 319 (1973).
28. I. Wender, *Catal. Rev.*, **14**, 97 (1976).
29. U. Tsao and J. W. Reilly, *Hydrocarbon Process.*, **57** (2), 133 (1978).
30. N. K. Kochar and R. L. Marcell, *Chem. Eng.* Jan. 28, 1980, p. 80.
31. D. W. Slocum, in *Catalysis in Organic Synthese*, (W. H. Jones, Ed.), Academic Press, New York, 1980, p. 245.
32. H. Bahrmann and B. Cornils, in *New Syntheses with Carbon Monoxide*, (J. Falbe, Ed.), Springer-Verlag, Berlin, 1980, p. 226.
33. J. Berty, L. Marko and D. Kallo, *Chem. Tech.* (Berlin), 8, 260 (1956).
34. T. Mizoroki and M. Nakayama, *Bull. Chem. Soc. Japan*, **37**, 236 (1964); **38**, 1876 (1965); **41**, 1628 (1968).
35. A. D. Riley and W. O. Bell, *US Patent* 3,248,432 (1966) to Commercial Solvents.
36. G. N. Butter, *US Patent* 3,285,948 (1966) to Commercial Solvents.
37. W. R. Pretzer, T. P. Kobylinski and J. E. Bozik, *US Patent* 4,133,966 (1979) to Gulf.
38. J. E. Bozik, T. P. Kobylinski and W. R. Pretzer, *US Patent* 4,239,924 (1980) to Gulf.
39. R. A. Fiato, *US Patent* 4,233,466 (1980) to Union Carbide.
40. L. H. Slaugh, *German Patent* 2,625,627 (1976) to Shell.
41. W. R. Pretzer and T. P. Kobylinski, *Ann. N. Y. Acad. Sci.*, **333**, 58 (1980).
42. Y. Sugi, K. Bando and Y. Takami, *Chem. Letters*, 63 (1981).
43. G. Braca, G. Sbrana, G. Valentini, G. Andrich and G. Gregorio, in *Fundamental Research in Homogeneous Catalysis*, Vol. 3, (M. Tsutsui, Ed.), Plenum Press, New York, 1979, p. 221.
44. G. S. Koermer and W. E. Slinkard, *Ind. Eng. Chem. Prod. Res. Dev.*, **17** (3), 231 (1978).
45. W. E. Slinkard and A. B. Baylis, *US Patent* 4,168,391 (1979) to Celanese.

46. W. E. Walker, *US Patent* 4,277,364 (1980) to Union Carbide.
47. T. E. Paxson, C. A. Reilly and D. R. Holecek, *J. Chem. Soc., Chem. Commun.*, 618 (1981).
48. H. Dumas, J. Levisalles and H. Rudler, *J. Organometal. Chem.*, 177, 239 (1979).
49. *C. and E. News*, April 7, 1980, p. 37.
50. R. F. Heck, *Advan. Organometal. Chem.*, 4, 243 (1966).
51. R. W. Goetz and M. Orchin, *J. Org. Chem.*, 27, 3698 (1962).
52. P. D. Taylor, *US Patent* 4,111,837 (1978) to Celanese.
53. W. R. Pretzer and T. P. Kobylinski, *US Patent* 4,239,705 (1980) to Gulf.
54. I. Wender, H. Greenfield, S. Metlin and M. Orchin, *J. Am. Chem. Soc.*, 74, 4079 (1952); see also I. Wender, H. Greenfield and M. Orchin, *J. Am. Chem. Soc.*, 73, 2656 (1951).
55. A. P. Gelbein, *ACS Div. Petrol. Chem. Preprints*, 26 (3), 749 (1981).
56. M. B. Sherwin and A. M. Brownstein, *British Patent* 2,007,652 (1979) to Chem. Systems.
57. M. B. Sherwin, A. M. Brownstein and J. Peress, *Eur. Patent Appl.* 4732 (1980) to Chem. Systems.
58. G. Braca, G. Sbrana, G. Valentini, G. Andrich and G. Gregorio, *J. Am. Chem. Soc.*, 100, 6238 (1978).
59. G. Braca, L. Paladini, G. Sbrana, G. Valentini, G. Andrich and G. Gregorio, *Ind.Œng. Chem. Prod. Res. Dev.*, 20, 115 (1981).
60. L. Gauthier and R. Perron, *Europ. Patent* 31,784 (1981) to Rhone-Poulenc.
61. *Res. Discl.* 208, 327 (1981) to Shell; *C.A.*, 95, 149909x (1981).
62. J. F. Knifton, *US Patent* 4,270,015 (1981) to Texaco.
63. J. F. Knifton, *Chem. Tech.*, 609 (1981).
64. J. F. Knifton, *J. Mol. Catal.*, 11, 91 (1981); *J. Chem. Soc. Chem. Commun.*, 41 (1981).
65. U. Romano, R. Tesei, M. M. Mauri and P. Rebora, *Ind. Eng. Chem. Prod. Res. Dev.*, 19, 396 (1980).
66. D. M. Fenton and P. J. Steinwand, *US Patent* 3,393,136 (1968).
67. K. Nishimura, S. Uchiumi, K. Fujii, K. Nishihira and H. Itatani, *ACS Div. Petrol. Chem. Preprints*, 24 (1), 355 (1979).
68. F. Rivetti and U. Romano, *J. Organometal. Chem.*, 154, 323 (1978).
69. J. E. Hallgren, G. M. Lucas and R. O. Mathews, *J. Organometal. Chem.*, 204, 135 (1981).
70. J. E. Hallgren and R. O. Mathews, *J. Organometal. Chem.*, 192, C12 (1980).
71. J. E. Hallgren and G. M. Lucas, *J. Organometal. Chem.*, 212, 135 (1981).

CHAPTER 8

NITROGEN-CONTAINING SYSTEMS

The incorporation of nitrogen-containing substrates into syn gas/methanol reactions increases considerably the number of interesting products that are potentially available from these systems. We have already mentioned in Chapter 6 that methanol and ammonia manufacture are usually integrated in one plant. Much of the hydrogen necessary for ammonia manufacture in the Haber process is produced by the Water Gas Shift reaction. Some of the carbon dioxide co-product from this reaction is used for the production of urea for fertilizer. The reactions involved are summarized below for an intregrated plant based on syn gas derived from methane steam reforming.

$$CH_4 + H_2O \longrightarrow CO + 3 H_2 \qquad (1)$$

$$CO + 2 H_2 \longrightarrow CH_3OH \qquad (2)$$

$$CO + H_2O \longrightarrow CO_2 + H_2 \qquad (3)$$

$$N_2 + 3 H_2 \longrightarrow 2 NH_3 \qquad (4)$$

$$2 NH_3 + CO_2 \longrightarrow H_2NCONH_2 + H_2O \qquad (5)$$

Ammonia and its simple derivatives such as amines, amides, etc. are, thus, potentially readily available in an integrated syn gas/methanol/ammonia plant for use in down-stream processes.

8.1 Existing Commercial Applications

The reaction of methanol with ammonia, at elevated temperatures and pressures, in the presence of a copper chromite catalyst, is used industrially for the manufacture of methylamines [1]:

$$NH_3 \xrightleftharpoons[H_2O]{CH_3OH} CH_3NH_2 \xrightleftharpoons[H_2O]{CH_3OH} (CH_3)_2NH \xrightleftharpoons[H_2O]{CH_3OH} (CH_3)_3N \qquad (6)$$

Yields are $ca.$ 94% on methanol and 97% on ammonia. Since the products are all in equilibrium, this reaction can be used for the manufacture of mono-, di- or trimethylamine by recycling the unwanted amines after fractionation. Currently much more dimethylamine is consumed than the other two.

The second major application is in the manufacture of hexamethylene tetramine by reaction of aqueous formaldehyde with ammonia at ambient temperature.

$$6 \, H_2CO + 4 \, NH_3 \rightleftharpoons (CH_2)_6N_4 + 6 \, H_2O \qquad (7)$$

8.2 Amide Formation

We have already noted in Chapter 4 that the Reppe carbonylation of olefins in the presence of ammonia or primary and secondary amines leads to the formation of amides (Reaction 8). Amide formation similarly results from the carbonylation of organic halides in the presence of ammonia or amines (Reaction 9)

$$RCH{=}CH_2 + CO + R'_2NH \xrightarrow{[Co_2(CO)_8]} RCH_2CH_2CONR'_2 \quad (8)$$

$$RX + CO + R'_2NH \xrightarrow{[Co_2(CO)_8]} RCONR'_2 + [HX] \qquad (9)$$

$$R' = H, \text{alkyl}; \quad X = \text{halide}$$

The reactions are carried out at elevated temperatures and pressures in the presence of metal carbonyl catalysts, notably $Co_2(CO)_8$. Reaction 9 has also been achieved under mild conditions ($35°C$, 1 bar) using pre-generated $NaCo(CO)_4$ as the catalyst [2]. The highly reactive carbonylation catalyst produced from the reaction of $Co(OAc)_2$ with $NaH/t\text{-AmONa}/CO$ can effect the conversion of aryl bromides to the corresponding amides at $60°C$ and atmospheric pressure [3]:

$$ArBr + CO + R_2NH \longrightarrow ArCONR_2 + [HBr] \qquad (10)$$

Similarly, a variety of aryl, heterocyclic and vinyl halides react with CO and primary or secondary amines, in the presence of a $(Ph_3P)_2PdCl_2$ catalyst at $60{-}100°C$ and atmospheric pressure, to give the corresponding amides [4].

Intramolecular equivalents of Reactions 8 and 9 are also known. The $Co_2(CO)_8$-catalysed carbonylation of unsaturated amines, for example, leads to the formation of lactams [5–7]:

$$CH_2{=}CHCH_2NH_2 + CO \longrightarrow \underset{\underset{H}{N}}{\text{[pyrrolidone ring]}}{=}O \qquad (11)$$

Similarly, N-alkylallylamines, which are readily available from allylchloride and primary amines or allylamine and alkyl chlorides, afford N-alkylpyrrolidones in good yield [7, 8]. This offers a potentially interesting route for the manufacture of the industrial solvent N-methylpyrrolidone, from allyl chloride, methylamine and carbon monoxide.

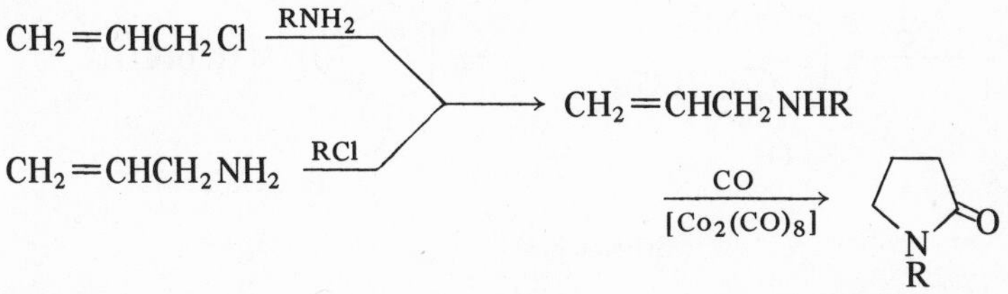

$$CH_2=CHCH_2Cl \xrightarrow{RNH_2}$$

$$CH_2=CHCH_2NH_2 \xrightarrow{RCl}$$

$$CH_2=CHCH_2NHR$$

$$\xrightarrow[{[Co_2(CO)_8]}]{CO} \quad \underset{\underset{R}{N}}{\bigvee}O \qquad (12)$$

These reactions can also be carried out in a single stage. For example, 2-pyrrolidone is formed by reacting allyl chloride, ammonia and CO in the presence of a cobalt chloride/Ph_3P catalyst at $250°C$ and 150 bar [9].

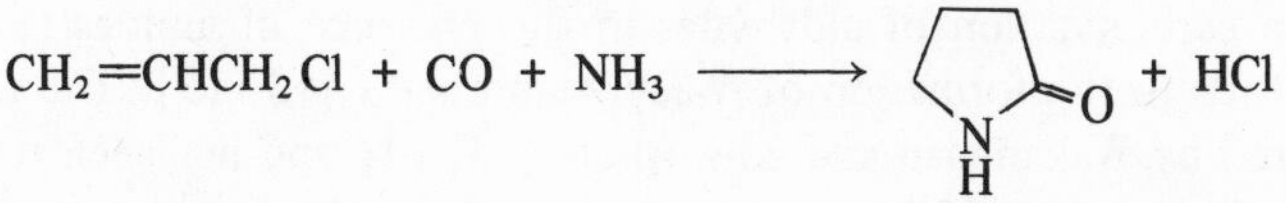

$$CH_2=CHCH_2Cl + CO + NH_3 \longrightarrow \underset{\underset{H}{N}}{\bigvee}O + HCl \qquad (13)$$

These reactions are not restricted to aliphatic amines as is illustrated by the following carbonylation of o-aminostyrene [9, 10].

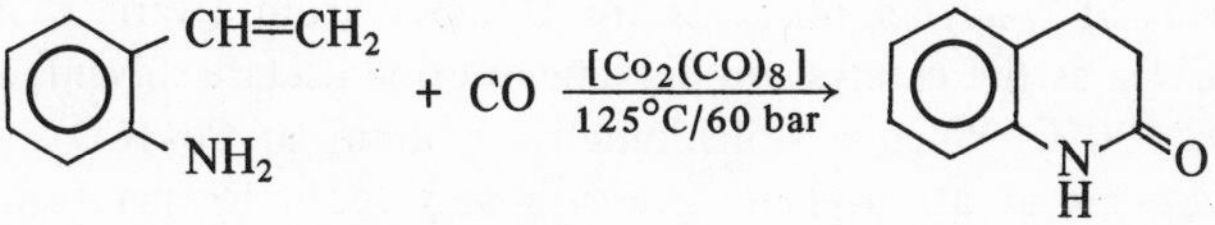

$$\underset{NH_2}{\overset{CH=CH_2}{\bigcirc}} + CO \xrightarrow[{125°C/60\ bar}]{[Co_2(CO)_8]} \underset{\underset{H}{N}}{\bigcirc}O \qquad (14)$$

The cobalt-catalysed carbonylation of unsaturated amides at $ca.$ $200°C$ and 300 bar affords cyclic imides [5, 6, 11]:

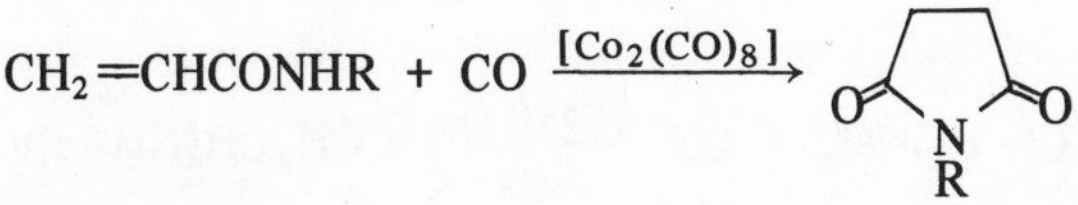

$$CH_2=CHCONHR + CO \xrightarrow{[Co_2(CO)_8]} O\underset{\underset{R}{N}}{\bigvee}O \qquad (15)$$

R = H, alkyl

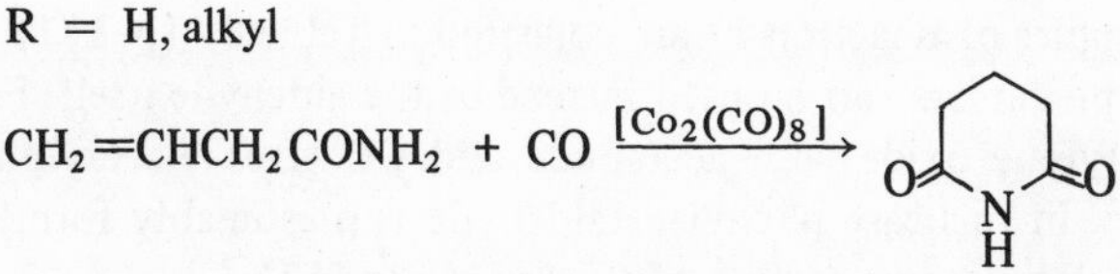

$$CH_2=CHCH_2CONH_2 + CO \xrightarrow{[Co_2(CO)_8]} O\underset{\underset{H}{N}}{\bigvee}O \qquad (16)$$

The mechanisms of these reactions are analogous to those of the inter-molecular carbonylations discussed in Chapter 4. The active catalyst is $HCo(CO)_4$, or the anion $Co(CO)_4^-$, and the reaction pathway involves alkyl- and acylcobalt intermediates as illustrated below (Scheme 8.1) for the formation of succinimide from acrylamide [11, 12].

$$CH_2{=}CHCONH_2 \; + \; HCo(CO)_4 \; \longrightarrow \; (CO)_4\,CoCH_2\,CH_2\,CONH_2$$

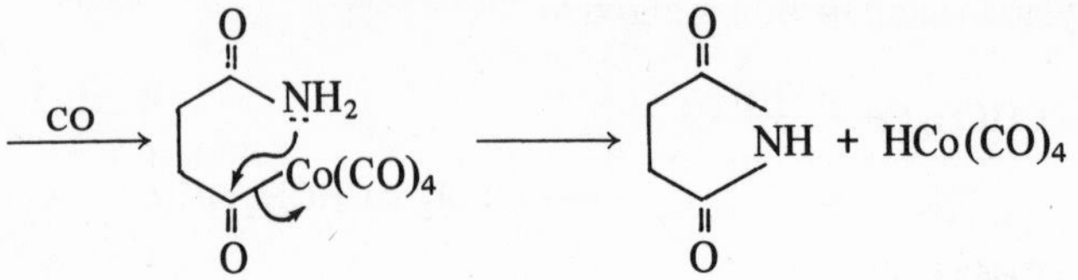

Scheme 8.1

8.3 Amine Acid Synthesis

Among the most interesting synthetic reactions of carbon monoxide is the catalytic carbonylation of aldehydes in the presence of amides (Reaction 17) which leads to the formation of N-acylamino acids [13–15]. The reaction was discovered by Wakamatsu and co-workers [13, 14] and has been referred to as *amidocarbonylation* [15].

$$RCHO + R'CONH_2 + CO \; \xrightarrow[H_2]{[Co_2(CO)_8]} \; RCH(NHCOR')CO_2H \qquad (17)$$

High yields of N-acylamino acids are obtained from a variety of aldehydes using $Co_2(CO)_8$ as the catalyst, in dioxane or ethyl acetate solvent, at 150–200 bar and *ca.* 100°C. When a stoichiometric amount of $Co_2(CO)_8$ is used the reaction proceeds at atmospheric pressure and room temperature [14]. The reaction is carried out in the presence of hydrogen $(CO/H_2 = 3/1)$ and the amide used is generally acetamide. For example, N-acetylalanine is produced in 94% yield by reaction of acetaldehyde with acetamide and CO at 120°C and 160 bar [13].

$$CH_3CHO + CH_3CONH_2 + CO \; \xrightarrow[H_2]{[Co_2(CO)_8]} \; CH_3CH(NHAc)CO_2H \quad (18)$$

Other examples of Reaction 17 are collected in Table 8.I (p. 171).

Aldehyde precursors can be used instead of the aldehyde itself. For example, reaction of styrene oxide with acetamide and CO gives N-acetylphenylalanine (Reaction 19). In this case phenylacetaldehyde is presumably formed *in situ* by $Co_2(CO)_8$-catalysed isomerisation of styrene oxide [13].

$$Ph\,\overset{\displaystyle O}{CH{-}\!\!\triangle\!\!{-}CH_2} + CH_3CONH_2 + CO \xrightarrow[H_2]{[Co_2(CO)_8]} PhCH_2CH(NHAc)CO_2H \quad (19)$$

Similarly, acetic anhydride can be used as a precursor of acetaldehyde via *in*

TABLE 8.I
Synthesis of N-acetylamino acids by amidocarbonylation[a]

$$RCHO + CH_3CONH_2 + CO \xrightarrow[H_2]{[Co_2(CO)_8]} RCH(NHAc)CO_2H$$

R	Solvent	RCH(NHAc)CO$_2$H Yield (%)	Reference
H	Dioxane	46	13
CH$_3$	Dioxane	94	13
CH$_3$(CH$_2$)$_2$	EtOAc	92–97	15
(CH$_3$)$_2$CH	Dioxane	70	13
PhCH$_2$	Dioxane	54	14
CH$_3$(CH$_2$)$_6$	EtOAc	84	15
CH$_3$(CH$_2$)$_8$	EtOAc	80	15
NC(CH$_2$)$_2$	Dioxane	58	14
CH$_3$S(CH$_2$)$_2$	EtOAc	64	14
CH$_3$O$_2$C(CH$_2$)$_2$	EtOAc	70	14

[a] Typical conditions: 70–120°C and 150–200 bar

situ hydrogenation. The amide component can also be replaced by a nitrile. For example, reaction of acetic anhydride with acetonitrile and syn gas gives, after hydrolysis, alanine [13]:

$$(CH_3CO)_2O + CH_3CN + CO/H_2 \xrightarrow[HOAc]{[Co_2(CO)_8]} CH_3CH(NHAc)CO_2H \quad (20)$$

$$\xrightarrow{H_2O/H^+} CH_3CH(NH_2)CO_2H$$

Since aldehydes are readily formed under the reaction conditions by $Co_2(CO)_8$-catalysed olefin hydroformylation, the aldehyde substrate can also be replaced by the appropriate olefin [16]. The N-acylamino acid is then formed according to the stoichiometry:

$$RCH{=}CH_2 + 2\,CO + RCONH_2 + H_2 \xrightarrow{[Co_2(CO)_8]}$$

$$RCH_2CH_2{-}CH\overset{\textstyle NHCOR}{\underset{\textstyle CO_2H}{\big\langle}} + RCH{-}CH\overset{\textstyle NHCOR}{\underset{\textstyle CO_2H}{\big\langle}} \quad (21)$$
$$\underset{\textstyle CH_3}{\phantom{RCH{-}CH}}$$

As in the hydroformylation reaction itself a mixture of linear and branched isomers is formed. For example, reaction of 1-butene with acetamide and syn gas at 100°C and 100 bar gave a 78% molar yield of N-acetylamino acids with a linear/branched ratio of 75/22:

$$CH_3CH_2CH{=}CH_2 + CH_3CONH_2 + 2\,CO + H_2$$

$$\xrightarrow[\text{dioxane}]{[CO_2(CO)_8]} CH_3(CH_2)_3{-}CH\begin{matrix}\diagup NHCOCH_3\\ \diagdown CO_2H\end{matrix} \quad +$$

(75%)

$$CH_3CH_2\underset{\displaystyle CH_3}{\underset{|}{C}H}{-}CH\begin{matrix}\diagup NHCOCH_3\\ \diagdown CO_2H\end{matrix} \qquad (22)$$

(22)%

1-Dodecene, on the other hand, reacts with a surprisingly high degree of linearity, the *N*-acetylamino acid being formed in 73% yield with 98% linearity:

$$CH_3(CH_2)_9CH{=}CH_2 + CH_3CONH_2 + 2\,CO + H_2$$

$$\xrightarrow[\text{dioxane}]{[Co_2(CO)_8]} CH_3(CH_2)_{11}CH\begin{matrix}\diagup NHCOCH_3\\ \diagdown CO_2H\end{matrix} \qquad (23)$$

Internal olefins also react and afford the linear acetylamino acid as the major product presumably via initial isomerisation of the olefin. Application of this chemistry to vinyl acetate and acrylonitrile affords methods for the synthesis of *dl*-threonine and *dl*-glutamic acid, respectively, as illustrated below.

$$CH_2{=}CHOAc + CH_3CONH_2 + 2\,CO + H_2$$

$$\xrightarrow[\text{2. hydrolysis}]{\text{1. }[Co_2(CO)_8]} CH_3CH(OH)CH(NH_2)CO_2H \qquad (24)$$

$$CH_2{=}CHCN + CH_3CONH_2 + 2\,CO + H_2$$

$$\xrightarrow[\text{2. hydrolysis}]{\text{1. }[Co_2(CO)_8]} HO_2CCH_2CH_2CH(NH_2)CO_2H \qquad (25)$$

The amidocarbonylation reaction would have even more synthetic utility if it were possible to obtain an asymmetric synthesis. Unfortunately, attempts to achieve asymmetric induction using chiral cobalt catalysts were unsuccessful [15]. The following mechanism satisfactorily explains the amidocarbonylation reaction (Scheme 8.2 p. 173).

Alternatively, by analogy with the cyclisation reactions described in the preceding section one might expect the α-acylamino acylcobalt intermediate (I) to undergo intramolecular acyl-cobalt cleavage to afford the oxazolone (II) as shown below. The latter then undergoes hydrolysis to give the *N*-acylamino acid.

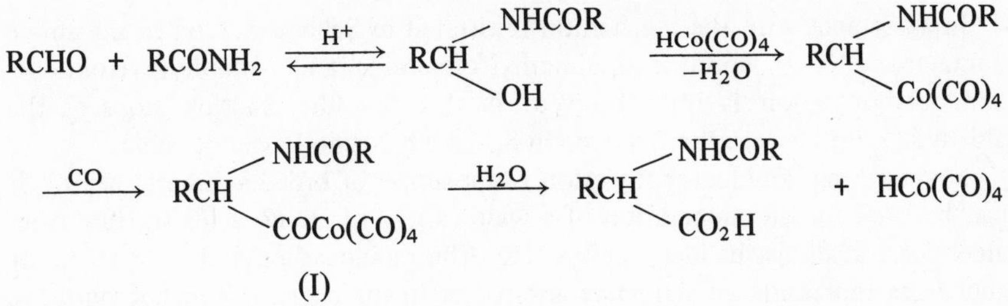

$$RCHO + RCONH_2 \; \underset{}{\overset{H^+}{\rightleftharpoons}} \; RCH{\overset{NHCOR}{\underset{OH}{\big<}}} \; \xrightarrow[-H_2O]{HCo(CO)_4} \; RCH{\overset{NHCOR}{\underset{Co(CO)_4}{\big<}}}$$

$$\xrightarrow{CO} \; RCH{\overset{NHCOR}{\underset{COCo(CO)_4}{\big<}}} \; \xrightarrow{H_2O} \; RCH{\overset{NHCOR}{\underset{CO_2H}{\big<}}} \; + \; HCo(CO)_4$$

(I)

Scheme 8.2

The isolation of an oxazolone when the reaction is carried out in the presence of molecular sieves lends support to this mechanism [17].

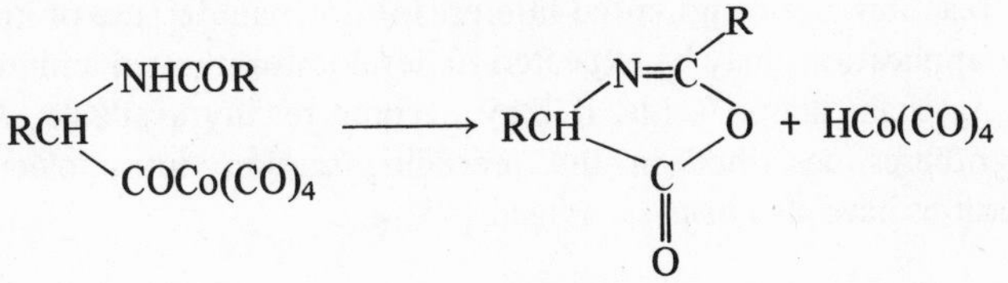

$$RCH{\overset{NHCOR}{\underset{COCo(CO)_4}{\big<}}} \; \longrightarrow \; RCH{\overset{N=C{\overset{R}{}}}{\underset{C{\overset{}{\underset{O}{\parallel}}}}{\big<}}O} \; + \; HCo(CO)_4 \qquad (26)$$

(II)

In a further modification of the amidocarbonylation reaction alcohols or organic halides are converted to N-acylamino acids by reaction with amides and syn gas in the presence of $Co_2(CO)_8$ as catalyst [18]. For example, N-acetyl-phenylalanine is formed in 77% yield by reaction of benzyl chloride with acetamide and syn gas in acetone at $120°C$ and 200 bar.

$$PhCH_2Cl + CH_3CONH_2 + 2CO + H_2$$

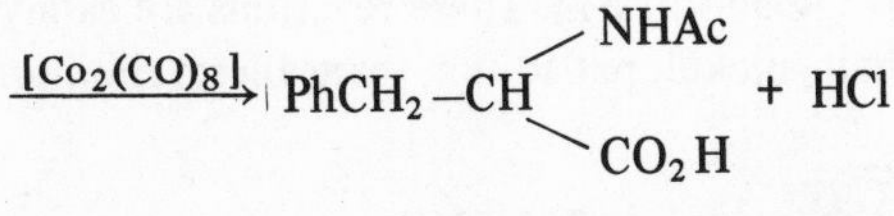

$$\xrightarrow{[Co_2(CO)_8]} PhCH_2-CH{\overset{NHAc}{\underset{CO_2H}{\big<}}} \; + \; HCl \qquad (27)$$

In these reactions the putative aldehyde intermediate is formed via hydro-carbonylation of the halide or alcohol as shown in Scheme 8.3. The aldehyde is subsequently converted to the N-acylamino acid via the pathway shown in Scheme 8.2.

$$R-X + HCo(CO)_4 \; \longrightarrow \; RCo(CO)_4 + HX$$

$$\xrightarrow{CO} \; RCOCo(CO)_4 \; \xrightarrow{H_2} \; RCHO + HCo(CO)_4$$

Scheme 8.3

In agreement with the mechanism illustrated in Scheme 8.2, when the amide contains no N–H bond, e.g. dimethylformamide and N-methylpyrrolidone, amidocarbonylation is not observed. In this case the reaction stops at the aldehyde stage because the first step in Scheme 8.2 is no longer possible.

Summarising, amidocarbonylation is a reaction of broad synthetic utility. It can be used for the preparation of a wide variety of amino acids starting from aldehydes, alcohols, halides, olefins, etc. The amidocarbonylation of aldehyde substrates represents an attractive alternative to the classical Strecker synthesis (Reaction 28).

$$RCHO + HCN + NH_3 \xrightarrow[-H_2O]{} RCH(NH_2)CN$$

$$\longrightarrow RCH(NH_2)CO_2H \qquad (28)$$

These reactions are of potential interest for the manufacture of known amino acids and applications may be expected to develop for unusual amino acids, such as the long chain amino acids, if they become readily available. As with the amide syntheses described in the preceding section, intramolecular amido-carbonylations have also been envisaged [13] e.g.

$$\qquad (29)$$

8.4 Amine Carbonylation

Catalytic carbonylation of primary and secondary amines leads, depending on the substrate and conditions, to the formation of the corresponding formamides (Reaction 30) or ureas (Reaction 31). These reactions are catalysed by carbonyl complexes of iron, cobalt, nickel, palladium, manganese, rhodium and ruthenium [19–25].

$$R_2NH + CO \longrightarrow R_2NCHO \qquad (30)$$

$$2\ RNH_2 + CO \longrightarrow RNHCONHR + H_2 \qquad (31)$$

For example, secondary aliphatic and heterocyclic amines, such as piperidine, are selectively converted to the corresponding N-formyl derivatives (Reaction 30) in the presence of $Co_2(CO)_8$ [20] or $Ru_3(CO)_{12}$ [21, 22] as catalysts. In the latter case the reaction proceeds under relatively mild conditions (75°C and 1 bar). The active catalyst in these reactions is probably the metal carbonylate anion, formed by reaction of the metal carbonyl with the amine. For example, $Co_2(CO)_8$ reacts with dimethylamine by the following disproportionation reaction [20]:

$$3 \, Co_2(CO)_8 + 20 \, Me_2NH$$

$$\longrightarrow 2 \, Co^{II}(Me_2NH)_6 \, [Co(CO)_4]_2 + 8 \, HCONMe_2 \qquad (32)$$

When a high pressure of CO is employed catalytic carbonylation is observed [20]. The key intermediate in these reactions is assumed to be a carbamoylmetal species (R_2NCOM). The observed increasing rate of carbonylation with increasing basicity of the amine is consistent with a mechanism involving rate-limiting attack of the amine on co-ordinated carbon monoxide. The formamide product is then formed by reductive elimination as shown in Scheme 8.4. The overall reaction is analogous to the formation of methyl formate through the reaction of methanol with carbon monoxide in the presence of a $Ru_3(CO)_{12}$ catalyst (see Chapter 7 section 5, p. 153).

$$M-CO + R_2NH \longrightarrow H-M-\overset{\overset{\displaystyle O}{\|}}{C}-NR_2 \longrightarrow R_2NCHO + M$$

Scheme 8.4

Primary amines generally give formamides or ureas depending on the conditions. For example, the carbonylation of n-butylamine in the presence of $[Rh(CO)_2Cl]_2$ as catalyst [23] at 160°C and 60 bar, affords a mixture of N-butylformamide and di-n-butylurea (Reaction 33). When the reaction is carried out in the presence of Me_3P (Me_3P : catalyst ratio = 6 : 1), on the other hand, N-butylformamide is formed in quantitative yield [23].

The selective carbonylation of primary aliphatic amines to the corresponding N-alkylformamides is also observed with $Fe(CO)_5$ as the catalyst at 200°C and 95 bar [24]. A mechanism involving a carbamoyliron intermediate was proposed to account for the results.

$$BuNH_2 + CO \xrightarrow{[Rh(CO)_2Cl]_2} BuNHCHO + BuNHCONHBu \qquad (33)$$
$$\phantom{BuNH_2 + CO \xrightarrow{[Rh(CO)_2Cl]_2}} 35\% \qquad\qquad 65\%$$

A plausible mechanism for urea formation involves β-hydrogen elimination from a carbamoylmetal species to give an isocyanate (Reaction 34). The latter then reacts with a molecule of amine to give the urea (Reaction 35). The catalytic cycle is completed by reaction of the amine and CO with the metal (carbonyl) hydride species formed in Reaction 34.

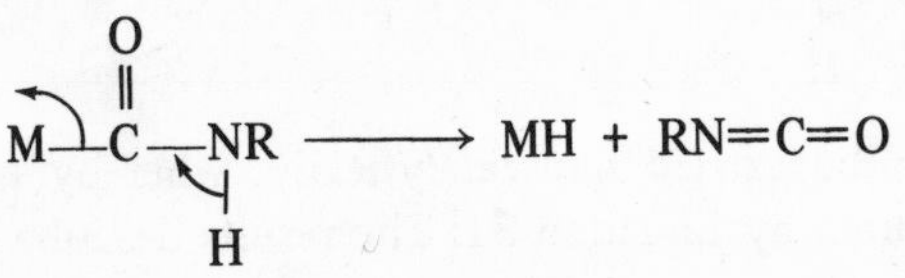

$$M-\overset{\overset{\displaystyle O}{\|}}{C}-NR \longrightarrow MH + RN{=}C{=}O \qquad (34)$$
$$\underset{H}{|}$$

$$RNCO + RNH_2 \longrightarrow RNHCONHR \qquad (35)$$

$$MH + RNH_2 + CO \longrightarrow M\overset{\overset{\displaystyle O}{\|}}{-}C-NHR + H_2 \qquad (36)$$

Reaction 34 is analogous to the formation of carbonyl compounds by β-hydrogen elimination from alkoxymetal compounds.

$$M-O-CR_2 \longrightarrow M-H + R_2C=O \qquad (37)$$

Dombek and Angelici [25] showed that $Mn_2(CO)_{10}$ reacts with primary aliphatic amines to give the carbamoylmanganese complex (III) which could be isolated and characterised. Under CO pressure (III) reacts to give the urea and $HMn(CO)_5$. It was suggested that this reaction proceeds via intermediate isocyanate formation as shown in Scheme 8.5.

$$Mn_2(CO)_{10} + 2\,RNH_2 \xrightarrow[- HMn(CO)_5]{} (CO)_4\,Mn\overset{\nearrow NH_2R}{\searrow_{CONHR}}$$

(III)

$$\longrightarrow (CO)_5\,Mn\overset{\nearrow \overset{\cdot\cdot}{N}H_2R}{\searrow_H} + RNCO \longrightarrow (CO)_5MnH + (RNH)_2CO$$

Scheme 8.5

Support for the intermediacy of isocyanates is provided by the observation that the reaction of amines with stoichiometric amounts of $PdCl_2$ in the presence of carbon monoxide results in the formation of isocyanates via *oxidative carbonylation* [26]

$$RNH_2 + CO + PdCl_2 \longrightarrow RNCO + Pd^0 + 2\,HCl \qquad (38)$$

The key step in Reaction 38 is presumably:

$$Cl-Pd^{II}-\overset{\overset{\displaystyle O}{\|}}{C}-NR \longrightarrow RNCO + Pd^0 + HCl \qquad (39)$$

The reaction can also be carried out catalytically, whereby hydrogen is co-produced [27] as required by Equation 31. The results described above for the rhodium system are also consistent with a mechanism involving the formation of an isocyanate intermediate via β-hydrogen elimination. Thus, it is well established that a prerequisite for facile β-hydrogen elimination from transition metal compounds is the presence of a vacant co-ordination site, i.e. co-ordinative unsaturation. Urea formation, if it involves Reaction 34, should therefore be suppressed by strongly co-ordinating ligands such as phosphines, as is observed.

Catalytic oxidative carbonylation is observed with CO/O_2 mixtures in the presence of a Pd^{II}/Cu^{II} catalyst combination, resulting in the formation of ureas, presumably via isocyanate intermediates [28].

$$2\,RNH_2\ +\ CO\ +\ \tfrac{1}{2}\,O_2\ \xrightarrow{\ [Pd^{II}/Cu^{II}]\ }\ RNHCONHR\ +\ H_2O \tag{40}$$

Oxidative carbonylation of primary amines is also observed with selenium catalysts [29] under relatively mild conditions (20°C and 1 bar). Interestingly, intramolecular carbonylations were also observed [29], under more forcing conditions (60°/50 bar):

$$H_2N(CH_2)_2NH_2\ +\ CO\ \xrightarrow[\tfrac{1}{2}O_2]{[Se]}\ HN\underset{\|}{\overset{}{\bigcirc}}NH\ +\ H_2O \tag{41}$$

98% yield

$$H_2N(CH_2)_3NH_2\ +\ CO\ \xrightarrow[\tfrac{1}{2}O_2]{[Se]}\ HN\underset{\|}{\overset{}{\bigcirc}}NH\ +\ H_2O \tag{42}$$

96% yield

The isolation of selenocarbamate salts as intermediates in these reactions strongly supports a mechanism involving the following steps [29]:

$$2\,RNH_2\ +\ Se\ +\ CO\ \longrightarrow\ RNH_3^+\ RNH{-}\overset{\overset{\textstyle O}{\|}}{C}{-}Se^- \tag{43}$$

$$RNH_3^+\ RNH{-}\overset{\overset{\textstyle O}{\|}}{C}{-}Se^-\ +\ O_2\ \longrightarrow\ (RNH)_2CO\ +\ Se\ +\ H_2O \tag{44}$$

The mechanism of the product forming step in this sequence has not been clearly resolved. It could involve isocyanate formation (pathway (a), Scheme 8.6) or attack of another amine molecule at the carbonyl group of the selenocarbamate (pathway b). Indeed, the question of β-hydrogen elimination *vs.* attack of

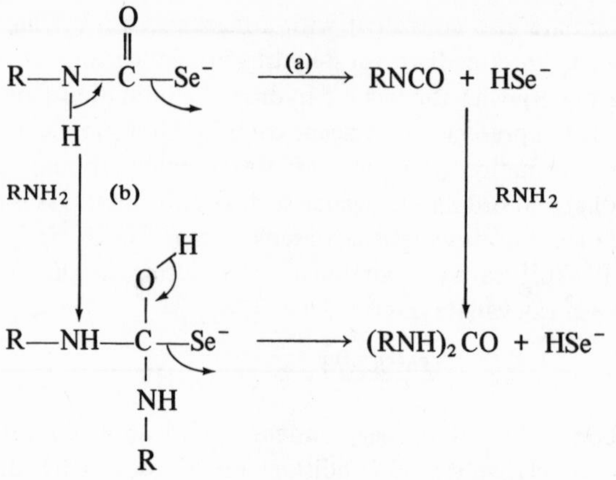

$$
\begin{array}{c}
\overset{\displaystyle O}{\overset{\|}{R-N-C-Se^-}} \xrightarrow{\ (a)\ } RNCO + HSe^- \\
\quad\;\; | \\
\quad\;\; H
\end{array}
$$

Scheme 8.6

RNH_2 at the carbonyl group is also applicable to the transition metal-catalysed urea formation discussed above.

8.5 Reactions of Nitro Compounds

The conversion of aromatic nitro compounds to the corresponding isocyanates is of considerable industrial interest since the latter are the raw materials for polyurethane manufacture. The classical method for carrying out this conversion is via catalytic hydrogenation and subsequent reaction with phosgene:

$$ArNO_2 + 3\,H_2 \longrightarrow ArNH_2 + 2\,H_2O \tag{45}$$

$$ArNH_2 + COCl_2 \longrightarrow ArNCO + 2\,HCl \tag{46}$$

Considerable research effort has been expended on finding new, phosgene-free routes to aromatic cyanates. One such process involves *reductive carbonylation* of the nitro compound by reaction with carbon monoxide (Reaction 47). This reaction proceeds in the presence of Group VIII metal catalysts, particularly palladium [30, 31] and rhodium [32] compounds at elevated temperatures and pressures.

$$ArNO_2 + 3\,CO \longrightarrow ArNCO + 2\,CO_2 \tag{47}$$

Alternatively, reductive carbonylation in the presence of alcohols affords the corresponding urethanes (Reaction 48), which are also suitable raw materials for polyurethane manufacture. The catalysts of choice for this reaction appear to be selenium [33, 34] or $PdCl_2$ [35] in the presence of tertiary amine co-catalysts.

$$ArNO_2 + 3\,CO + ROH \longrightarrow ArNHCO_2R + 2\,CO_2 \qquad (48)$$

Good results were obtained, for example, using selenium in the presence of strongly basic amines at 160°C and 70 bar [33]. Under these conditions 2,4-dinitrotoluene was converted to 2,4-*bis*(ethylcarbamoyl)toluene in 80–89% yield (Reaction 49). The product is widely used as a raw material for polyurethane manufacture.

The reaction of aromatic nitro compounds with carbon monoxide/water mixtures in alkaline media in the presence of metal carbonyl catalysts, notably $Fe(CO)_5$ [36], $Fe_3(CO)_{12}$ [37] and $Ru_3(CO)_{12}$ [37], results in reduction to the corresponding amines according to the stoichiometry:

$$ArNO_2 + 3\,CO + H_2O \longrightarrow ArNH_2 + 3\,CO_2 \qquad (50)$$

This represents an alternative to catalytic hydrogenation for carrying out this transformation. Process improvements have been achieved by the application of phase transfer catalysis to these systems [38,39]. For example, using $Ru_3(CO)_{12}$ as the metal catalyst, and benzyltriethylammonium chloride as the phase transfer catalyst in an aqueous NaOH/organic solvent system, amines were obtained in excellent yields (see Table 8.II) at room temperature and atmospheric pressure [39]. The reaction was also applicable to aliphatic nitro compounds.

The active catalysts in these systems are metal carbonyl hydride anions, formed by attack of hydroxide ion at co-ordinated CO followed by elimination of CO_2:

$$M_3(CO)_{12} + R_4NOH \longrightarrow R_4N^+[M_3(CO)_{11}CO_2H]^-$$
$$\xrightarrow[-CO_2]{} R_4N^+[HM_3(CO)_{11}]^- \qquad (51)$$

The reduction of aromatic nitro compounds with CO/H_2O mixtures, to give the corresponding amines, is also catalysed by the rhodium cluster carbonyl, $Rh_6(CO)_{16}$, in the presence of tertiary amines [40, 41]. An interesting modification of this reaction is the $Rh_6(CO)_{16}$-catalysed formation of Schiff bases by reaction of aromatic nitro compounds with benzaldehyde and CO in tertiary amine solvents [42]:

$$ArNO_2 + PhCHO + 3\,CO \xrightarrow[[Rh_6(CO)_{16}]]{R_3N} ArN{=}CHPh + 3\,CO_2 \qquad (52)$$

TABLE 8.II

$Ru_3(CO)_{12}$-catalysed reduction of nitro compounds to
amines under phase transfer conditions [39]

$$RNO_2 + 3CO + H_2O \xrightarrow[Ru_3(CO)_{12},\ PhCH_2NEt_3Cl]{NaOH/C_6H_6/MeOCH_2CH_2OH} RNH_2 + 2CO_2$$

Nitro compound	Reaction time (hr)	Product	Yield (%)
Nitrobenzene	9	Aniline	100
p-Nitrotoluene	8	p-Toluidine	94
p-Nitroanisole	7	p-Anisidine	84
p-Chloronitrobenzene	3	p-Chloroaniline	100
p-Nitrobenzophenone	4.5	p-Aminobenzophenone	100
2,6-Dimethylnitrobenzene	25	2,6-Dimethylaniline	8
2-Nitrofluorene	20	2-Fluorenylamine	95
1-Nitropropane	17	n-Propylamine	85

Conditions: $Ru_3(CO)_{12}$ (0.03 mmol), $PhCH_2NEt_3Cl$ (0.25 mmol), RNO_2 (5 mmol), CO (1 bar), 5N NaOH (10 mL), benzene (25 mL) and 2-methoxyethanol (5 mL) at room temperature.

The mechanism of reduction of nitro compounds by metal carbonyls and carbonylate anions has not been clearly resolved but probably involves successive oxygen transfer reactions between the nitro compound and co-ordinated carbon monoxide. This can be conveniently envisaged as formally involving cycloaddition of RNO_2 and RNO to a metal-carbonyl double bond and subsequent extrusion of carbon dioxide (see Scheme 8.7).

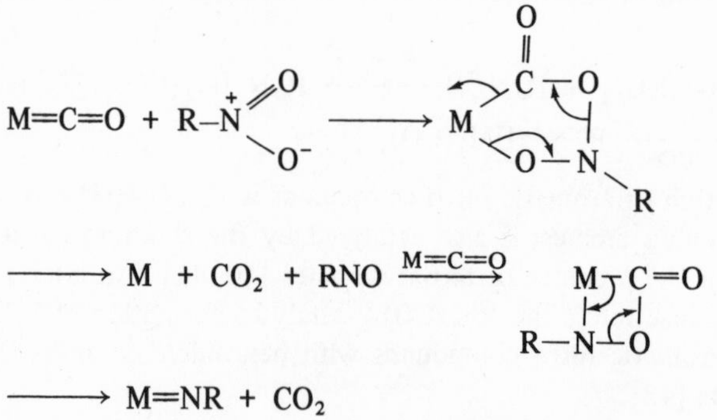

Scheme 8.7

Conversion of the alkylimidometal (M=NR) complex to the observed products proceeds via the reactions outlined in Scheme 8.8.

$$
M=NR \xrightarrow{-M}
\begin{cases}
\xrightarrow{CO} RNCO \\[1ex]
\xrightarrow{CO/R'OH} RNHCO_2R' \\[1ex]
\xrightarrow{CO/H_2O} RNH_2 + CO_2 \\[1ex]
\xrightarrow{CO/R'CHO} RN=CHR' + CO_2
\end{cases}
$$

Scheme 8.8

Alternative oxygen transfer pathways can be envisaged whereby oxometal species are formed (Reaction 53). In order that *catalytic* reduction may take place the oxometal species must be subsequently reduced by (co-ordinated) carbon monoxide (Reaction 54). This probably accounts for the fact that these reductions are not always catalytic in the metal carbonyl [37].

$$M + RNO_2 \longrightarrow M=O + RNO \tag{53}$$

$$M=O + CO \longrightarrow M + CO_2 \tag{54}$$

8.6 Tertiary Amine Synthesis

We have already mentioned in Chapter 4 that tertiary amines are produced in good yield by olefin hydroformamination (Reaction 55), which is catalysed by cobalt and rhodium complexes under hydroformylation conditions.

$$RCH=CH_2 + CO + 2H_2 + R'_2NH \longrightarrow R(CH_2)_3NR'_2 + H_2O \tag{55}$$

The reaction, which is sometimes referred to as aminomethylation, can also be carried out using CO/H_2O mixtures and metal carbonyl catalysts under alkaline conditions [43–45]:

$$RCH=CH_2 + 3CO + H_2O + R'_2NH \longrightarrow R(CH_2)_3NR'_2 + 2CO_2 \tag{56}$$

The mixed metal combinations Rh/Fe [43, 44] and Ru/Fe [44] were shown to be particularly effective catalysts for Reaction 56. Both Reactions 55 and 56 involve aldehyde intermediates, formed via olefin hydroformylation. An alternative approach is to use the aldehyde as the substrate. Thus, phosphine-modified cobalt carbonyl complexes catalyse the N-alkylation of morpholine by carbonyl compounds according to the stoichiometry [46]:

$$O{\overset{\frown}{\underset{\smile}{}}}NH + {\overset{R}{\underset{R'}{\searrow}}}C{=}O + CO \xrightarrow{H_2O} O{\overset{\frown}{\underset{\smile}{}}}N{-}CHRR' + CO_2 \quad (57)$$

For example, excellent yields of N-methyl-, N-isopropyl- and N-benzyl-morpholine were obtained using $Co_2(CO)_8$ in conjunction with 1,2-*bis*(diphenyl-phosphino)ethane at 120°C and 100 bar [46]. This transformation constitutes an alternative to conventional reductive amination of carbonyl compounds with amines and hydrogen in the presence of hydrogenation catalysts.

8.7 Nitrile Formation

The direct conversion of mixtures of syn gas and ammonia to acetonitrile, over supported iron or molybdenum catalysts at 500°C (Reaction 58), has been described [47]. Selectivities in the range 40–45% were reported at *ca.* 40% conversion, the major by-products being methane and carbon dioxide.

$$2 CO + 2 H_2 + NH_3 \longrightarrow CH_3CN + H_2O \quad (58)$$

We presume, by analogy, that methanol could also be converted to acetonitrile by the sequence:

$$CH_3OH + CO + NH_3 \xrightarrow{-H_2O} CH_3CONH_2 \longrightarrow CH_3CN + H_2O \quad (59)$$

To our knowledge this reaction has not yet been reported but should, in principle, be catalysed by methanol carbonylation catalysts in the presence of a dehydrating agent. We have mentioned earlier (Chapter 6) that acetonitrile can be converted to acrylonitrile by condensation with formaldehyde. Reaction 59 could, therefore, form the basis for a methanol-based route to acrylonitrile via the overall stoichiometry:

$$2 CH_3OH + NH_3 + CO \longrightarrow CH_2{=}CHCN + H_2O + H_2 \quad (60)$$

8.8 Summary

The reactions of CO or syn gas in combination with organic nitrogen-containing compounds, in the presence of metal carbonyl catalysts, afford a broad array of synthetically useful reactions. Most, if not all, of them have largely gone unnoticed by organic chemists. In many cases they form useful alternatives to classical, multi-step procedures involving the use of metal cyanides or phosgene. Although the examples described in the literature generally involve elevated temperatures and/or pressures it is worth noting that the latest carbonylation techniques described in Chapter 4 have not been applied to these systems. We expect, therefore, that once organic chemists begin to appreciate their potential these reactions will be widely applied.

References

1. *Hydrocarbon Process.*, 58(11), 185 (1979).
2. R. F. Heck and D. S. Breslow, *J. Am. Chem. Soc.*, 85, 2779 (1969).
3. J. J. Brunet, C. Sidot, B. Loubinoux and P. Caubere, *J. Org. Chem.*, 44, 2199 (1979).
4. A. Schoenberg and R. F. Heck, *J. Org. Chem.*, 39, 3327 (1974).
5. A. Mullen, in *New Syntheses with Carbon Monoxide,* (J. Falbe. Ed.), Springer-Verlag, Berlin, 1980, p. 414.
6. J. Falbe, *Carbon Monoxide in Organic Synthesis*, Springer-Verlag, Berlin, 1970, p. 146. .
7. J. Falbe and F. Korte, *Chem. Ber.*, 98, 1928 (1965).
8. J. Falbe, H. Weitkamp and F. Korte, *Tetrahedron Letters*, 2677 (1965).
9. R. F. Prince, *US Patent* 3,637,743 (1968) to Atlantic Richfield.
10. G. Soder, *German Patent* 1,620,391 (1966); see also ref. 5.
11. J. Falbe and F. Korte, *Chem. Ber.*, 95, 2680 (1962).
12. J. Falbe, *Angew. Chem.*, 78, 532 (1966).
13. H. Wakamatsu, J. Uda and N. Yamagami, *US Patent* 3,766,266 (1973) to Ajinomoto.
14. H. Wakamatsu, J. Uda and N. Yamakami, *J. Chem. Soc. Chem. Commun.*, 1540 (1971).
15. J. J. Parnaud, G. Campari and P. Pino, *J. Mol. Catal.*, 6, 341 (1979).
16. R. Stern, A. Hirschauer, D. Commereuc and Y. Chauvin, *British Patent* 2,000,132 (1978) to Institut Français du Pétrole.
17. K. Izawa, T. Yamashita and Y. Ozawa, *Abstr. 26th Int. Congr. Pure Appl. Chem.*, Tokyo, 1977, Sect. Ia, p. 80.
18. T. Yukawa, N. Yamakami, M. Honma, Y. Komachiya and H. Wakamatsu, *German Patent* 2,364,039 (1974) to Ajinomoto.
19. A. Rosenthal and I. Wender, in *Organic Syntheses via Metal Carbonyls*, (I. Wender and P. Pino, Eds.) Vol. 1, Wiley Interscience, New York, 1968, p. 405.
20. H. W. Sternberg, I. Wender, R. A. Friedel and M. Orchin, *J. Am. Chem. Soc.*, 75, 3148 (1953); W. Hieber, J. Sedlmeier and W. Abeck, *Chem. Ber.*, 86, 700 (1953).
21. G. L. Rempel, W. K. Teo, B. R. James and D. V. Plackett, *Advan. Chem. Ser.*, 132, 166 (1974).
22. J. J. Byerley, G. L. Rempel, N. Takabe and B. R. James, *J. Chem. Soc. Chem. Commun.*, 1482 (1971).
23. D. Durand and C. Lassau, *Tetrahedron Letters*, 2329 (1971).
24. B. D. Dombek and R. J. Angelici, *J. Catal.*, 48, 433 (1977).
25. B. D. Dombek and R. J. Angelici, *J. Organometal. Chem.*, 134, 203 (1977).
26. E. W. Stern and M. L. Spector, *J. Org. Chem.*, 26, 3126 (1961).
27. J. Tsuji and N. Iwamoto, *J. Chem. Soc. Chem. Commun.*, 380 (1966).
28. Y. L. Sheludyakov, V. A. Golodov and D. V. Sokolskii, *Dokl. Akad. Nauk SSSR*, 249 (3), 658 (1979).
29. N. Sonada, T. Yasuhara, K. Kondo, T. Ikeda and S. Tsutsumi, *J. Am. Chem. Soc.*, 93, 6344 (1971).
30. P. D. Hammond, W. M. Clarke and W. I. Denton, *US Patent* 3,832,372 (1974); P. D. Hammond, J. A. Scott, W. M. Clarke and W. I. Denton, *US Patent* 3,828,089 (1974); P. D. Hammond and N. B. Franco, *US Patent* 3,823,174 (1974); P. D. Hammond and J. A. Scott, *US Patent* 3,812,169 (1974) all to Olin Corporation.
31. W. W. Pritchard, *US Patent* 3,576,836 (1971) to Du Pont.
32. W. B. Hardy and R. P. Bennett, *Tetrahedron Letters*, 961 (1967).
33. *British Patent* 1,485, 108 (1977) to Mitsui Toatsu.
34. J. G. Zajacek, J. J. McCoy and K. E. Fuger, *US Patent* 3,895,054 (1975) to Atlantic Richfield.

35. J. G. Zajacek and J. J. McCoy, *US Patent* 3,993,685 (1976) to Atlantic Richfield.
36. K. Cann, T. Cole, W. Slegeir and R. Pettit, *J. Am. Chem. Soc.*, **100**, 3969 (1978).
37. H. Alper and K. E. Hashem, *J. Am. Chem. Soc.*, **103**, 6514 (1981); see also T. Okano, K. Fujiwara, H. Konishi and J. Kiji, *Chem. Letters*, 1083 (1981); J. M. Landesberg, L. Katz and C. Olsen, *J. Org. Chem.*, **37**, 930 (1972).
38. H. des Abbayes and H. Alper, *J. Am. Chem. Soc.*, **99**, 98 (1977).
39. H. Alper and S. Amaratunga, *Tetrahedron Letters*, **21**, 2603 (1980).
40. R. Ryan, G. M. Wilemon, M. P. Dalsanto and C. Pittman, *J. Mol. Catal.*, **5**, 319 (1979).
41. A. F. M. Iqbal, *Tetrahedron Letters*, 3385 (1971).
42. A. F. M. Iqbal, *J. Org. Chem.*, **37**, 2791 (1972).
43. A. F. M. Iqbal, *Helv. Chim. Acta*, **54**, 1440 (1971).
44. R. M. Laine, *J. Org. Chem.*, **45**, 3370 (1980).
45. F. Jachimowicz and J. W. Raksis, *J. Org. Chem.*, **47**, 445 (1982).
46. Y. Sugi, A. Matsuda, K. Bando and K. Murata, *Chem. Letters*, 363 (1979).
47. S. Olivé and G. Olivé, *German Patent* 2,629,189 (1977) to Monsanto.

ADDITIONAL READING

A. Mullen, 'Ring Closure Reactions with Carbon Monoxide' in *New Syntheses with Carbon Monoxide*, (J. Falbe, Ed.), Springer-Verlag, Berlin, 1980, p. 414.
A. F. M. Iqbal, 'Deoxygenate Organic Compounds with Carbon Monoxide', *Chem. Tech.*, **4**, 566 (1974)

CHAPTER 9

DIRECT CONVERSION OF SYN GAS TO OXYGENATES

In the preceding chapters we have discussed two approaches to the synthesis of oxygen-containing industrial chemicals from syn gas. The first approach involves the conversion of syn gas to lower olefins followed by conventional technology for further transformations to oxygenated derivatives. The second involves methanol synthesis and subsequent reactions of methanol with carbon monoxide or syn gas. In this chapter we shall discuss a third approach, namely the direct conversion of syn gas to C_2 and higher oxygenates.

The overall transformations leading to methanol (Reaction 1) and various C_2 oxygenates (Reactions 2–5) are outlined below.

$$CO + 2\,H_2 \rightleftharpoons CH_3OH \tag{1}$$

$$2\,CO + 4\,H_2 \rightleftharpoons CH_3CH_2OH + H_2O \tag{2}$$

$$2\,CO + 3\,H_2 \rightleftharpoons CH_3CHO + H_2O \tag{3}$$

$$2\,CO + 2\,H_2 \rightleftharpoons CH_3CO_2H \tag{4}$$

$$2\,CO + 3\,H_2 \rightleftharpoons HOCH_2CH_2OH \tag{5}$$

Since all of these reactions involve a decrease in the number of moles an increase in pressure leads to an increase in conversion. However, the susceptibility of conversion to increasing pressure varies significantly for the different reactions. Ethanol, acetaldehyde and acetic acid show very high equilibrium conversions ($> 90\%$) at pressures above 10 bar. Methanol, on the other hand, shows high equilibrium conversions only at pressures exceeding *ca.* 200 bar. For high equilibrium conversions to ethylene glycol even higher pressures are required. Furthermore, a comparison of the free energies of formation of the oxygenates with those of low molecular weight alkanes reveals that the latter are the thermodynamically favourable products at all pressures and temperatures. It is evident, therefore, that the synthesis of C_2 oxygenates requires kinetic control in order to minimise alkane formation.

9.1 Lower Alcohol Synthesis

Methanol is the only alcohol currently produced on a commercial scale by direct

185

reaction of carbon monoxide with hydrogen. Synthetic ethanol is produced by the hydration of ethylene. Similarly, the higher secondary alcohols isopropanol and secondary butanol are manufactured via the hydration of propylene and butenes, respectively. The manufacture of the straight chain alcohols n-propanol and n-butanol, on the other hand, involves the hydroformylation of ethylene and propylene, respectively (see Chapter 4).

A comparison of the prices of these alcohols reveals that methanol, which is entirely based on syn gas, is significantly cheaper than the other alcohols which are derived from olefin feedstocks. It is not surprising, therefore, that considerable industrial research effort has been focussed on syn gas routes to lower (C_2-C_4) alcohols.

Since many transition metal complexes catalyse both methanol synthesis (Reaction 1) and methanol homologation (Reaction 6) they should, in principle, be capable of catalysing the direct conversion of syn gas to ethanol (Reaction 2) and higher alcohols. Basically two approaches have been used to achieve this end, by employing the carbonylation catalysts cobalt and rhodium or by employing modified methanol synthesis catalysts.

$$CH_3OH + CO + 2H_2 \longrightarrow CH_3CH_2OH + H_2O \qquad (6)$$

COBALT AND RHODIUM CATALYSTS

An example of the use of a typical carbonylation/homologation catalyst is the reported conversion of syn gas to ethanol, in the presence of a $Co_2(CO)_8$ catalyst in dioxane solvent, at $182°C$ and 238 bar [1, 2]. Selectivities to ethanol were, however, rather low ($ca.$ 20–25%). The major product was ethylene glycol (25–45%) and by-products included methanol, n-propanol, n-butanol, propylene glycol, formate esters, methane and carbon dioxide. These results have been interpreted on the basis of a mechanism (Scheme 9.1) involving formaldehyde as a common intermediate [3].

The primary products are methanol, methyl formate and ethylene glycol. Virtually all of the other products are derived from the methanol via HCoCO)$_4$-catalysed homologation (Scheme 9.2), involving acetaldehyde as a common intermediate.

In contrast to the low selectivities obtained in dioxane, the $Co_2(CO)_8$-catalysed reaction of syn gas in 1,2-dimethoxyethane (glyme) and related polyglycol ethers, under roughly the same conditions, was reported to produce ethanol in selectivities up to 80% [4]. However, a more recent investigation has demonstrated that most of the ethanol is derived from the solvent and only $ca.$ 5% from the direct reaction of CO and H_2 [5]. The formation of ethanol from methoxy-terminated polyglycol solvents probably involves initial cleavage by the

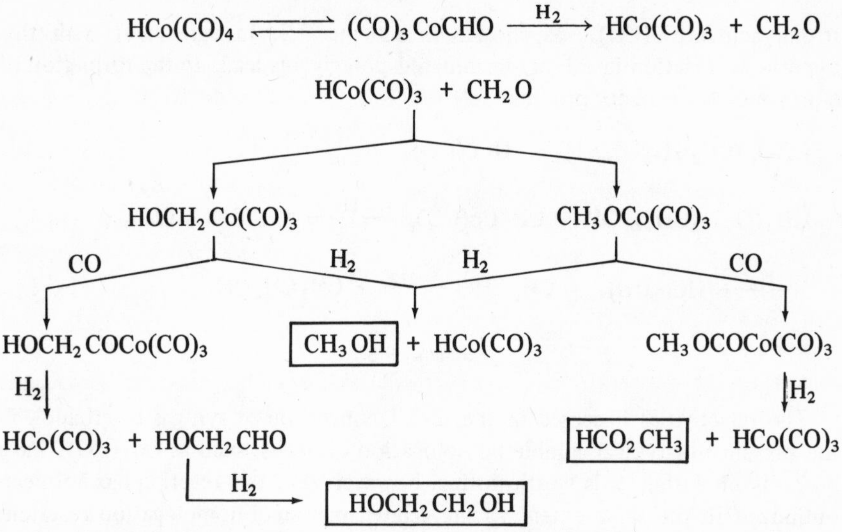

$$HCo(CO)_4 \rightleftharpoons (CO)_3 CoCHO \xrightarrow{H_2} HCo(CO)_3 + CH_2O$$

$$HCo(CO)_3 + CH_2O$$

HOCH$_2$Co(CO)$_3$ CH$_3$OCo(CO)$_3$

CO   H$_2$ H$_2$ CO

HOCH$_2$COCo(CO)$_3$ $\boxed{CH_3OH}$ + HCo(CO)$_3$ CH$_3$OCOCo(CO)$_3$

H$_2$ $\Big\Vert$H$_2$

HCo(CO)$_3$ + HOCH$_2$CHO $\boxed{HCO_2CH_3}$ + HCo(CO)$_3$

$\xrightarrow{H_2}$ $\boxed{HOCH_2CH_2OH}$

Scheme 9.1

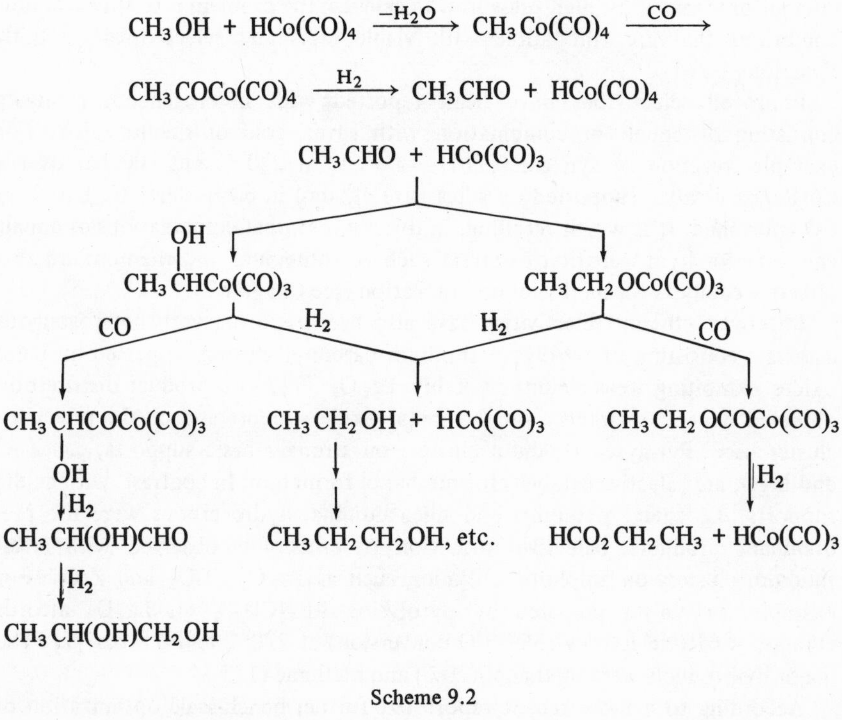

$$CH_3OH + HCo(CO)_4 \xrightarrow{-H_2O} CH_3Co(CO)_4 \xrightarrow{CO}$$

$$CH_3COCo(CO)_4 \xrightarrow{H_2} CH_3CHO + HCo(CO)_4$$

$$CH_3CHO + HCo(CO)_3$$

OH
CH$_3$CHCo(CO)$_3$ CH$_3$CH$_2$OCo(CO)$_3$

CO   H$_2$ H$_2$ CO

CH$_3$CHCOCo(CO)$_3$ CH$_3$CH$_2$OH + HCo(CO)$_3$ CH$_3$CH$_2$OCOCo(CO)$_3$
|
OH $\Big\Vert$H$_2$
$\downarrow$H$_2$
CH$_3$CH(OH)CHO   CH$_3$CH$_2$CH$_2$OH, etc.   HCO$_2$CH$_2$CH$_3$ + HCo(CO)$_3$

$\downarrow$H$_2$
CH$_3$CH(OH)CH$_2$OH

Scheme 9.2

strong acid, $HCo(CO)_4$, as shown in Scheme 9.3. In agreement with this hypothesis, reaction in ethoxy-terminated polyglycols leads to the formation of *n*-propanol as the major product.

$$CH_3O(CH_2CH_2O)_nCH_3 + HCo(CO)_4 \longrightarrow$$

$$CH_3(OCH_2CH_2)_nOH + CH_3Co(CO)_4 \xrightarrow{CO} CH_3COCo(CO)_4$$

$$\xrightarrow{H_2} HCo(CO)_4 + CH_3CHO \xrightarrow{H_2} CH_3CH_2OH$$

Scheme 9.3

The selectivities observed in the direct conversion of syn gas to ethanol, in the presence of typical soluble homologation catalysts, such as $Co_2(CO)_8$ have been disappointing. It is worth noting, however, that this reaction has not been optimized to the same extent as the related methanol homologation reaction. Indeed, most studies have been concerned with optimizing the yield of ethylene glycol. The formation of the latter is favoured, with respect to methanol (the ethanol precursor), by high pressures. In essence, the problem is to find reaction conditions that are compatible with viable rates and selectivities for both Reactions 1 and 6.

Improved selectivities have been reported with heterogeneous catalysts consisting of cobalt in combination with silver, gold or rhenium [6]. For example, reaction of syn gas ($CO/H_2 = 1 : 1$) at 280°C and 100 bar over a Co/Re/Ba catalyst supported on silica gave ethanol in 60% selectivity based on CO consumed. It is worth recalling, in this context, that combinations of cobalt and superior hydrogenation catalysts such as ruthenium and rhenium are also effective catalysts for methanol homologation (see Chapter 7).

Improved ethanol selectivities have also been reported with heterogeneous catalysts consisting of pyrolysed rhodium carbonyl clusters dispersed on basic oxides exhibiting weak acidity, notably La_2O_3 [7]. The product distribution observed depends markedly on the metal oxide support and on the precursor cluster used. Pyrolysed rhodium clusters on strongly basic supports, e.g. ZnO and MgO, are selective catalysts for methanol formation. In contrast, with acidic supports, e.g. silica, γ-alumina and silica-alumina, hydrocarbons were the predominant products. Selective formation of ethanol is observed with small rhodium clusters on amphoteric oxides, such as La_2O_3, TiO_2 and ZrO_2. For example, a catalyst prepared by pyrolysing $Rh_4(CO)_{12}$ on La_2O_3 affords ethanol in 61% selectivity (36% CO conversion) at 220°C and 0.8 bar [7]. The major by-products were methanol (20%) and methane (12%).

According to a more recent report [8] further bench-scale optimization of the $Rh_4(CO)_{12}/La_2O_3$ system (Sagami process) has resulted in ethanol selectivities of 75–80% for a one-pass CO conversion level of 8–10% at 250–290°C and 50 bar. Moreover, the catalyst showed no significant loss of activity after one month of operation. This process would seem to offer important advantages over a two-step route via methanol since the operating pressure of 50 bar is the same pressure at which syn gas is produced.

C_2 oxygenates are also the major products of the reaction of syn gas over a Rh/Fe mixed metal catalyst supported on silica [9]. Major C_2 products were ethanol and acetic acid but ethanol selectivities (up to *ca.* 50%) were significantly less than those reported for the pyrolysed rhodium clusters discussed above. This catalyst may be more suitable for direct conversion of syn gas to acetic acid which, by an appropriate choice of conditions, can be formed in selectivities up to 60%. In this context it may be recalled that rhodium is the catalyst of choice for methanol carbonylation to acetic acid (see Chapter 7).

MODIFIED METHANOL SYNTHESIS CATALYSTS

It has been known for some time that modification of a zinc chromite methanol synthesis catalyst with alkali metal salts leads to the formation of substantial amounts of isobutanol [10].

$$CO + H_2 \xrightarrow[K_2O]{ZnCr_2O_4} CH_3OH + (CH_3)_2CHCH_2OH, \text{etc.} \qquad (7)$$

The yield of normal, straight-chain alcohols is quite low and the 'isosynthesis' reaction is not suitable for producing ethanol and *n*-propanol. A typical product distribution at 380–450°C and 300–400 bar is 50% methanol, 20–40% isobutanol with the remainder consisting of ethanol and higher alcohols.

Workers at the Institut Français du Pétrole showed, in 1978, that a catalyst consisting of a mixture of Cu and Co oxides and alkali metal salts together with the oxide of either Cr, Fe, V or Mn is effective for the direct conversion of syn gas to C_2–C_4 normal alcohols [10, 11]. The IFP process affords a selectivity to alcohols exceeding 95% at 35% conversion per pass when operated at 250°C and 60 bar. A typical alcohol product distribution (wt. %) is 20% methanol, 38% ethanol, 21% *n*-propanol and 17% *n*-butanol.

$$CO + H_2 \xrightarrow{IFP \ catalyst} CH_3OH + CH_3CH_2OH + CH_3CH_2CH_2OH, \text{etc.} \qquad (8)$$

The mechanism of the IFP process has not been discussed but the catalyst clearly contains both methanol synthesis (Cu, Cr) and methanol homologation (Co) components. We assume, therefore, that it involves classical methanol synthesis followed by cobalt-mediated homologation. The process would appear to offer an attractive method for the synthesis of normal C_2–C_4 alcohols.

However, it has been pointed out that the economics of the process (and analogous processes) are less attractive than anticipated, mainly due to the stoichiometry of the reaction [10]. Thus, in the direct synthesis of ethanol from syn gas ca. 30% wt. of the feedstock is converted to water. This contrasts with methanol (and ethylene glycol) synthesis in which all of the syn gas can theoretically be converted to product.

9.2 Conversion to Ethylene Glycol

The first demonstration of ethylene glycol synthesis from syn gas was reported by Gresham in 1953 [12]. It was shown than cobalt carbonyl-catalysed hydrogenation of CO at pressures up to 1000 bar gave a mixture of monofunctional oxygenates (methanol, ethanol, acetic acid, methyl acetate, etc.). In contrast, reaction at 1500–5000 bar afforded increasing amounts of dihydric and polyhydric alcohols, including ethylene glycol and glycerol and their derivatives. The selectivity of glycol formation could be increased using inert solvents or solvents that react with the glycol products.

Soluble cobalt salts, e.g. cobalt acetate, are suitable catalyst precursors (the active catalyst is presumably $HCo(CO)_4$). For example, the conversion of syn gas at 225–250°C/3000 bar in the presence of cobalt acetate catalyst and acetic acid solvent gave a mixture of ethylene glycol diacetate and glycerol triacetate as the major products [12]. On employing an aqueous cobalt acetate solution, free ethylene glycol together with its mono- and di-formate esters are formed. The yield of glycols is influenced markedly by the pressure, glycols and monofunctional oxygenates being obtained in a 1 : 3 ratio and 1 : 1 ratio at 1500 bar and 3000 bar, respectively. It is noteworthy in this context that according to a more recent report [1] substantial yields of ethylene glycol (up to 45%) are formed in the cobalt-catalysed hydrogenation of CO in dioxane solvent at significantly lower pressures (238 bar). Considering the conflicting nature of these two sets of results it would seem necessary to establish, unequivocally, that the ethylene glycol is not derived from the dioxane solvent.

As mentioned earlier methanol and ethylene glycol are probably formed from a common hydroxymethylcobalt intermediate, via competing hydrogenation and CO insertion processes (Scheme 9.4).

$$HCo(CO)_4 + H_2 \longrightarrow HOCH_2Co(CO)_3 \begin{array}{c} \xrightarrow{H_2} HCo(CO)_3 + CH_3OH \\ \xrightarrow{CO} HOCH_2COCo(CO)_3 \end{array}$$

$$HOCH_2COCo(CO)_3 \xrightarrow{H_2} HOCH_2CHO \xrightarrow{H_2} HOCH_2CH_2OH$$

Scheme 9.4

Subsequently, Union Carbide workers showed that soluble rhodium complexes catalyse the formation of ethylene glycol from syn gas at elevated pressures [13–17]. The Union Carbide process employs a homogeneous rhodium carbonyl catalyst which is generated *in situ* from a variety of catalyst precursors, e.g. rhodium salts, carbonyls and acetylacetonates. The reaction affords a mixture consisting mainly of methanol and ethylene glycol, together with smaller amounts of propylene glycol and glycerol. The conditions are severe, typically 210–250°C and 500–3000 bar but selectivities to ethylene glycol of up to *ca.* 70% have been claimed. The absence of methanol carbonylation and homologation products, such as acetic acid and ethanol, is expected since soluble rhodium complexes are not effective catalysts for these reactions in the absence of iodide promotors.

$$CO + H_2 \xrightarrow{\text{Rh catalyst}} CH_3OH + HOCH_2CH_2OH$$

(major products)

$$+ CH_3CH(OH)CH_2OH + HOCH_2CH(OH)CH_2OH \tag{9}$$

(minor products)

The exact nature of the catalytically active species has not been unequivocally established but it is believed to be the rhodium carbonyl cluster anion, $[Rh_{12}(CO)_{34}]^{2-}$, formed through the equilibria shown below (Reaction 10). The infrared spectra of reaction mixtures under typical operating conditions show the presence of both $[Rh(CO)_4]^-$ and $[Rh_{12}(CO)_{34}]^{2-}$ in varying concentrations [17, 18].

$$2 [Rh_6(CO)_{15}H]^- \underset{H_2}{\overset{-H_2}{\rightleftharpoons}} [Rh_{12}(CO)_{30}]^{2-} \underset{-4 CO}{\overset{4 CO}{\rightleftharpoons}} [Rh_{12}(CO)_{34}]^{2-} \tag{10}$$

The rate and selectivity of the reaction is influenced by the temperature and pressure, the nature of the solvent and the addition of promotors. The preferred solvents are those which exhibit good cation solvation properties, e.g. sulfolane and tetraglyme, $CH_3O(CH_2CH_2O)_3CH_3$, or even molten crown ethers such as 18-crown-6 [17]. The selectivity to ethylene glycol and the catalytic activity are increased by the addition of promotors such as alkali metal salts and/or organic nitrogen and oxygen Lewis bases. The alkali metal cations form the counter ions for the rhodium carbonyl anion. The solvent effect can be understood in terms of activity enhancement via the formation of reactive, "naked" carbonyl anions in solvents that decrease the tendency for ion pairing.

High pressures are required to provide for high selectivities to ethylene glycol since competing methanol formation is favoured at low pressures. Due to the relatively low activity of the catalyst (*ca.* 10^4 times less active than the rhodium-catalysed hydroformylation of propylene under comparable conditions), high temperatures are needed to achieve reasonable conversion rates, thus necessitating extremely high pressures. Pressures of 400–600 bar appear to be the lower limit for an adequate conversion rate, pressures of 1500–2000 bar being preferred. The severe reaction conditions present a serious obstacle to commercialisation of the Union Carbide process and there is an obvious need for more active catalysts. However, judging by the amount of effort which has already been devoted to this system it would seem unlikely that further major improvements will be forthcoming.

Iridium [19] and ruthenium [20] carbonyl complexes were also reported to catalyse the formation of ethylene glycol from syn gas, in the presence of organic bases such as 2-hydroxypyridine. These reactions also require severe conditions of temperature and pressure and activities are lower than with the rhodium system.

More recent studies by Dombek [21–23] and by Knifton [24, 25] have shown that ruthenium complexes are, under certain conditions, able to catalyse ethylene glycol formation at lower pressures. In inert solvents ruthenium complexes catalyse the selective formation of methanol from syn gas at 230°C and 340 bar [23]. In contrast, when the reaction is performed in the presence of acetic acid, usually as the solvent, methyl acetate is formed as the major product together with smaller amounts of ethylene glycol diacetate and ethyl acetate [21, 23, 24]. Similar reactions are observed using other carboxylic acids instead of acetic acid.

Infrared studies demonstrated that the catalyst is present mainly as $Ru(CO)_5$ irrespective of the catalyst precursor used [23]. The mechanism shown in Scheme 9.5 was postulated to account for the effect of carboxylic acids on the course of the reaction.

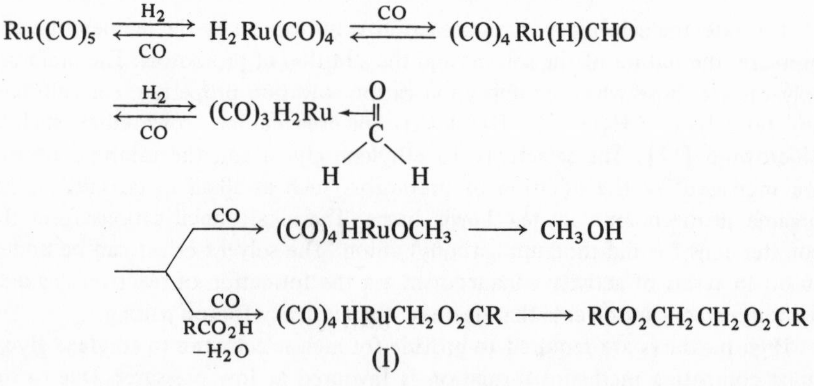

$$Ru(CO)_5 \underset{CO}{\overset{H_2}{\rightleftharpoons}} H_2Ru(CO)_4 \underset{}{\overset{CO}{\rightleftharpoons}} (CO)_4Ru(H)CHO$$

$$\underset{CO}{\overset{H_2}{\rightleftharpoons}} (CO)_3H_2Ru \longleftarrow \overset{O}{\underset{C}{\parallel}} \overset{}{\underset{H \quad H}{\diagdown}}$$

$$\overset{CO}{\longrightarrow} (CO)_4HRuOCH_3 \longrightarrow CH_3OH$$

$$\underset{RCO_2H \atop -H_2O}{\overset{CO}{\longrightarrow}} (CO)_4HRuCH_2O_2CR \longrightarrow RCO_2CH_2CH_2O_2CR$$

$$(I)$$

Scheme 9.5

A key feature of the mechanism is the intermediacy of an acyloxymethyl-ruthenium complex (I). Support for this mechanism was gained from a study [26] of the reactions of the model acyloxymethylmanganese carbonyl complex (II). Reaction of (II) with hydrogen at 7 bar and 75°C gave ethylene glycol ester in good yield (see Scheme 9.6).

$$(CO)_5 MnCH_2 O_2 CBu^t \rightleftharpoons (CO)_4 MnCOCH_2 O_2 CBu^t$$

(II)

$$\longrightarrow HMn(CO)_4 + Bu^t CO_2 CH_2 CHO \longrightarrow Bu^t CO_2 CH_2 CH_2 OH$$

Scheme 9.6

A plausible alternative to the mechanism outlined in Scheme 9.5 involves competitive acyloxylation and hydrogenation of a hydroxymethylruthenium intermediate as shown in Scheme 9.7.

$$(CO)_4 HRuCHO \xrightarrow{H_2} (CO)_4 HRuCH_2 OH$$

$$(CO)_4 HRuCH_2 OH \bigg\langle \begin{array}{l} \xrightarrow{H_2} H_2 Ru(CO)_4 + CH_3 OH \\ \\ \xrightarrow{RCO_2 H} (CO)_4 HRuCH_2 O_2 CR \end{array}$$

$$(CO)_4 HRuCH_2 O_2 CR \xrightarrow{CO} (CO)_4 HRuCOCH_2 O_2 CR, \text{ etc.}$$

Scheme 9.7

Ruthenium complexes have been shown to catalyse ethylene glycol formation in the presence of iodide promotors [22] or in molten quaternary ammonium or phosphonium salts, such as tetrabutylphosphonium bromide [25]. Thus, reaction at 200–250°C and 400–1000 bar using a $Ru_3(CO)_{12}/KI$ catalyst combination, in cation complexing solvents such as sulfolane, afforded methanol and ethylene glycol, in a ratio of ca. 5 : 1, as the primary products [22].

Ruthenium-catalysed reaction in a tetrabutylphosphonium bromide melt at 220°C and 430 bar gave ethylene glycol and ethylene glycol monoalkyl ethers in yields up to 30%, with other major products being methanol and ethanol [25]. The ruthenium hydrocarbonyl cluster $[HRu_3(CO)_{11}]^-$, was implicated as the active catalyst (c.f. the rhodium-catalysed reaction discussed earlier). Glycol formation was suggested to involve a hydroxymethylruthenium complex as a key intermediate as shown in Scheme 9.8.

$$\text{HRu(CO)} \xrightarrow{\text{H}_2} \text{RuCH}_2\text{OH} \xrightarrow{\text{CO}} \text{RuCOCH}_2\text{OH}$$

$$\xrightarrow{\text{H}_2} \text{RuCH(OH)CH}_2\text{OH} \xrightarrow{\text{H}_2} \text{HOCH}_2\text{CH}_2\text{OH}$$

Scheme 9.8

Summarising, the direct conversion of syn gas to ethylene glycol is catalysed by soluble complexes of Group VIII metals, notably cobalt, rhodium and ruthenium. However, the severe reaction conditions and/or poor selectivities represent serious drawbacks and tend to make the indirect routes discussed in Chapters 6 and 7 more attractive commercially. Because of its enormous potential, however, considerable industrial effort continues to be devoted to the direct route.

9.3 Conversion to Lower Carboxylic Acids

Rhodium-based catalysts are known to be effective for both the conversion of syn gas to methanol (see above) and the carbonylation of methanol to acetic acid (see Chapter 7). Rhodium catalysts should, therefore, be capable of effecting the direct conversion of syn gas to acetic acid. Indeed, this appears to be the case, as Union Carbide workers have reported the direct conversion of syn gas to C_2 oxygenates, consisting mainly of ethanol and acetic acid, over rhodium based catalysts [9]. The catalyst consists of rhodium in combination with other metals, such as iron, supported on silica. However, the selectivities to C_2 oxygenates are poor to moderate, being accompanied by significant amounts of hydrocarbon formation. We conclude, therefore, that this direct route does not present a serious challenge to the well-established indirect route via methanol synthesis and methanol carbonylation.

As mentioned in Chapter 7, soluble ruthenium complexes, in the presence of iodide promotors, catalyse the homologation of acetic acid to lower carboxylic acids. They are not effective, however, for the conversion of syn gas to acetic acid, the preferred pathway being ethylene glycol formation (see preceding section). Based on these results a promising approach to the direct conversion of syn gas to lower carboxylic acids could be via the use of mixed rhodium/ruthenium catalysts in combination with iodide promotors.

9.4 Summary

The direct conversion of syn gas to oxygenates, such as ethylene glycol, acetic acid, ethanol and other lower alcohols, appears to be feasible. However, with perhaps the notable exception of the selective conversion to ethanol over pyrolysed rhodium cluster catalysts (Sagami process), these processes are characterized by poor selectivities and/or severe reaction conditions.

In this context it should also be mentioned that modified Fischer–Tropsch catalysts are also able to effect the direct conversion of syn gas to oxygenates, with primary alcohols being the major products [27]. However, as in the Fischer–Tropsch reaction itself, the modified version affords products with a wide range of molecular weights. Consequently, the value of this reaction as a source of industrial chemicals is questionable.

References

1. H. M. Feder and J. W. Rathke, *Ann. N. Y. Acad. Sci.*, **333**, 45 (1980).
2. J. W. Rathke and H. M. Feder, *J. Am. Chem. Soc.*, **100**, 3623 (1978).
3. D. R. Fahey, *J. Am. Chem. Soc.*, **103**, 136 (1981).
4. R. J. Daroda, J. R. Blackborow and G. Wilkinson, *J. Chem. Soc. Chem. Commun.*, 1098 (1980).
5. T. E. Paxson, C. A. Reilly and D. R. Holecek, *J. Chem. Soc. Chem. Commun.*, 618 (1981).
6. H. Hachenberg, F. Wunder, E. I. Leupold and H. J. Schmidt, *Eur. Patent Appl.* 21,330 (1981) to Hoechst; *C.A.* **94**, 174317q (1981).
7. M. Ichikawa, *J. Chem. Soc. Chem. Commun.*, 566 (1978).
8. *Chem. Econ. Eng. Rev.*, **11**(5), 15 (1979).
9. P. C. Ellgen and M. M. Bhasin, *US Patent* 4,014,913 and 4,162,262 (1977) to Union Carbide.
10. G. M. Intille, *Preprints* Div. Petrol. Chem., Am. Chem. Soc. Meeting Honolulu, April 1979, p. 318; and references cited therein.
11. A. Sugier and E. Freund, *German Patent* 2,748,097 (1978) to Institut Français du Pétrole.
12. W. F. Gresham, *US Patent* 2,636,046 (1953) to Du Pont.
13. R. L. Pruett and W. E. Walker, *US Patent* 3,833,634 (1974) and 3,957,857 (1976) to Union Carbide.
14. R. L. Pruett, *Ann. N. Y. Acad. Sci.*, **295**, 239 (1977).
15. W. E. Walker, E. S. Brown and R. L. Pruett, *US Patent* 3,878,214 and 3,878,292 (1975) to Union Carbide.
16. W. E. Walker and J. B. Cropley, *US Patent*, 3,940,432 (1976) to Union Carbide.
17. L. Kaplan, *US Patent*, 4,162,261 (1979) to Union Carbide.
18. J. L. Vidal and W. E. Walker, *Inorg. Chem.*, **19**, 896 (1980).
19. R. C. Williamson and T. P. Kobylinski, *US Patent* 4,170,606 (1979) to Gulf Research and Development.
20. R. C. Williamson and T. P. Kobylinski, *US Patent* 4,170,605 (1979) to Gulf Research and Development.
21. B. D. Dombek, *J. Am. Chem. Soc.*, **102**, 6857 (1980).
22. B. D. Dombek, *J. Am. Chem. Soc.*, **103**, 6508 (1981).
23. B. D. Dombek, in *Catalytic Activation of Carbon Monoxide,* (P. C. Ford, Ed.), *ACS Symp. Series*, **152**, 213 (1981).
24. J. F. Knifton, *J. Chem. Soc. Chem. Commun.*, 188 (1981).
25. J. F. Knifton, *J. Am. Chem. Soc.*, **103**, 3959 (1981).

26. B. D. Dombek, *J. Am. Chem. Soc.*, **101**, 6466 (1979).
27. B. Cornils and W. Rottig, in *Chemical Feedstocks from Coal*, (J. Falbe, Ed.), Wiley, New York, 1982, p. 467.

CHAPTER 10

SUMMARY – DIRECTIONS FOR THE FUTURE

Since the 1950s the most important feedstocks for the organic chemicals industry have been derived mainly from oil. Lower olefins such as ethylene, propylene, butenes and butadiene and the aromatics benzene, toluene and xylenes constitute the key base chemicals in this industry. This dependence on a single raw material for both liquid fuels and chemicals places the chemical industry in an extremely vulnerable position. This vulnerability has become increasingly evident in recent years and has led to a burgeoning research effort on alternative sources of industrial chemicals.

An attractive feature of syn gas as a raw material for chemicals manufacture is the fact that it is readily available from a wide variety of sources. It can be produced from virtually any carbon source, including natural gas, coal and biomass. This flexibility makes chemicals manufacture from syn gas a viable proposition in a variety of geographical situations. Thus, irrespective of whether natural gas, coal or agricultural wastes are in good supply at a particular time and location, syn gas technology can be applied to chemicals production. It offers, for example, the possibility of using otherwise useless organic wastes, alone or in combination with coal, as raw materials for chemicals.

10.1 Strategic Options

There are essentially two points of divergence in a chemicals from syn gas strategy (Scheme 10.1). Since the syn gas could be produced in various geographical locations, on-site conversion to the readily transportable base chemical methanol would seem to give an attractive degree of flexibility. If necessary, the methanol could subsequently, at the point of use, be partially reconverted to syn gas.

The options at the second point of divergence – methanol conversion to lower olefins *vs*. direct conversion to industrial chemicals – are not mutually exclusive. Conversion to lower olefins has the advantage that further conversion can be via existing petrochemical processes, sometimes involving reaction with carbon monoxide (carbonylation) or syn gas (hydroformylation). On the other hand, direct conversion of methanol to oxygen-containing derivatives avoids

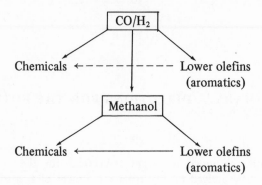

Scheme 10.1

the substantial weight loss inherent in hydrocarbon formation.

Conversion of methanol to lower olefins can be via direct "cracking" over zeolite catalysts or via homologation to ethanol and dehydration. An attractive feature of the former method is the possibility of integration with liquid fuels production (this also applies to the direct conversion of syn gas to lower olefins). The latter method (homologation) is attractive because ethanol, in addition to being converted to ethylene, is also a convenient starting material for other chemicals, e.g. acetaldehyde, butadiene. Moreover, there is overlap with another alternative source of ethanol: the fermentation of biomass. The various options are summarized in Scheme 10.2.

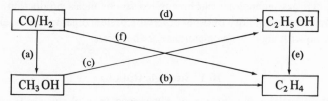

(a) Methanol synthesis   (d) Ethanol synthesis
(b) Methanol cracking   (e.g. Sagami process)
(c) Methanol homologation   (e) Ethanol dehydration
                                 (f) Ethylene synthesis
                                      (e.g. modified FT)

Scheme 10.2

10.2 Scope and Limitations

Methanol constitutes a versatile base chemical which can be converted to a wide range of important industrial chemicals by reaction with carbon monoxide or syn gas. Some involve processes that have already been commercialized, e.g. acetic acid via methanol carbonylation, whilst others are in various stages of development. A number of these products are illustrated in Scheme 10.3. Many of them are in turn versatile intermediates which can be converted to important downstream derivatives.

CH3CO2H
EtOH
(CH3CO)2O
CH3CHO
CH3CO2Et
H2CO
CH3OH
CH2=CHOAc
(CO2CH3)2
CH3CH2CO2H
(CH3O)2CO
H3CO2C(CH2)4CO2CH3
RCH2CH2CO2CH3

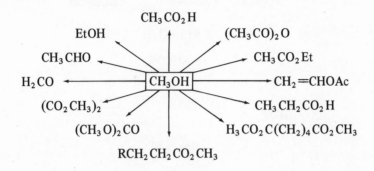

Scheme 10.3

One of these downstream derivatives, ethylene glycol, deserves special mention, partly because of its industrial importance and partly because of the many approaches which have been conceived for its synthesis. In addition to these possibilities, outlined in Scheme 10.4, the feasibility of the direct conversion of syn gas to ethylene glycol has also been demonstrated.

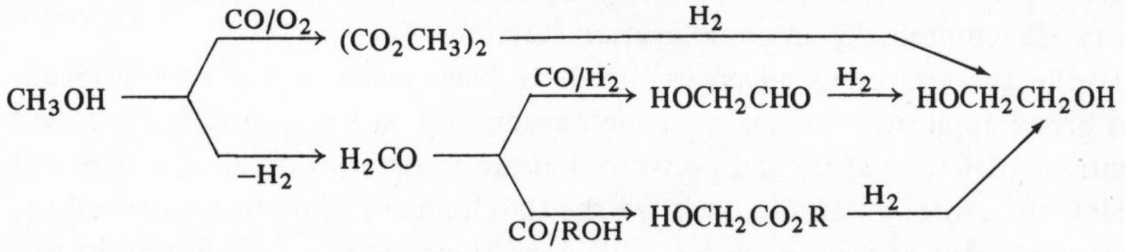

CO/O2
(CO2CH3)2
H2
CH3OH
CO/H2
HOCH2CHO
H2
HOCH2CH2OH
H2CO
–H2
H2
HOCH2CO2R
CO/ROH

Scheme 10.4

The reactions of carbon monoxide and syn gas are not only relevant to the manufacture of bulk chemicals. They have also been successfully applied to the synthesis of more complicated organic molecules. For example, the reactions of carbon monoxide or syn gas with olefins and organic halides have broad synthetic utility for the manufacture of fine chemicals. In particular, the reactions with organic halides (see Scheme 10.5) often proceed under relatively mild conditions with high selectivities.

Most of the reactions outlined in Schemes 10.3–5 have one feature in

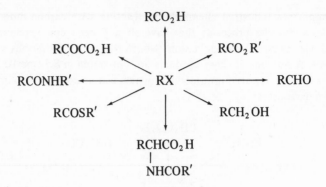

R= alkyl, allyl, benzyl, aryl, heterocycle
X= halide

Scheme 10.5

common; they are catalysed by complexes of Group VIII metals, notably cobalt, rhodium and ruthenium. The active catalytic species is generally the metal carbonyl hydride or carbonylate anion. Cobalt carbonyl, $Co_2(CO)_8$, was one of the first catalysts to be employed in these reactions and is still widely applied. However, in some reactions it is gradually being replaced by the more active rhodium-based catalysts. Ruthenium carbonyls have not received much attention until fairly recently but appear to show considerable promise as versatile catalysts, particularly in reactions where both carbonylation and hydrogenation activity are required, e.g. in homologation reactions.

Despite the enormous advances that have been made in this area in recent years broad application of many of these reactions is still impeded by the severe conditions of temperature and pressure that are sometimes required. There still remains, therefore, a definite need for the development of more active catalysts which allow for operation under mild conditions. It is worth noting in this context that the rate-limiting step in many carbonylation and related processes is the generation of the active catalyst by ligand dissociation. At the relatively high temperatures that are required for a reasonable rate of ligand dissociation the catalyst is often stable only at high carbon monoxide pressures. Techniques for low temperature activation would, therefore, also solve the problem of high pressures in many cases.

We have already mentioned in Chapter 2 that light-induced ligand dissociation constitutes a potentially useful technique for low temperature catalyst generation. A new development is the recently reported photochemical generation of active ruthenium carbonyl species isolated on high surface area silica [1].

Immobilization of the catalyst precursor circumvents a frequently encountered problem in such systems, namely that active, co-ordinatively unsaturated species, formed via photoactivation, react with each other to form inactive clusters. Another interesting development is the use of electrocatalysis in ligand substitution processes [2, 3].

Recent improvements in carbonylation techniques include the use of phase transfer catalysis and the generation of highly active catalysts through the use of strong bases such as NaH/RONa (see Chapter 5). The application of such techniques is gradually bringing these reactions into the realm of mainstream organic synthesis. These techniques are particularly relevant to the synthesis of fine chemicals and we expect that they will be widely applied in the near future.

The scope of carbonylation and related reactions in organic synthesis would be further expanded by the successful application of asymmetric catalysis. Although catalytic asymmetric hydrogenation is a well-developed technology, the optical yields observed in asymmetric carbonylations have, up till now, been disappointingly low.

Improved techniques for product separation and catalyst recovery and recycle are also areas which merit further attention. The application of phase transfer catalysis (see sbove) is an important development in this context. Further development of catalyst immobilization techniques may also be expected to lead to more efficient catalyst recovery and recycle.

10.3 Mechanism

The fundamental steps in carbonylations and related processes are, thanks to the substantial effort devoted in recent years to mechanistic studies, fairly well-defined. These are the fundamental processes that are ubiquitous in organometallic chemistry: ligand replacement and activation, oxidative addition, migratory insertion and reductive elimination. The mechanistic details of these fundamental processes are, however, not completely resolved. Indeed, there does not appear to be one all-encompassing mechanism to account for each type of process.

This is not so surprising. In classical organic chemistry fundamental processes, such as nucleophilic substitution, proceed by several different mechanisms, depending on the nature of the substrate, solvent and nucleophile. Likewise, organometallic processes involve different mechanisms depending on the nature of the substrate, metal, ligands and solvent. The fundamental question to be resolved in many organometallic processes is whether the initial step involves a one- or two-electron transfer. These alternatives are illustrated in Scheme 10.6 for the oxidative addition of an alkyl halide to a metal complex.

When the initial step is a one-electron transfer this can result in a radical non-

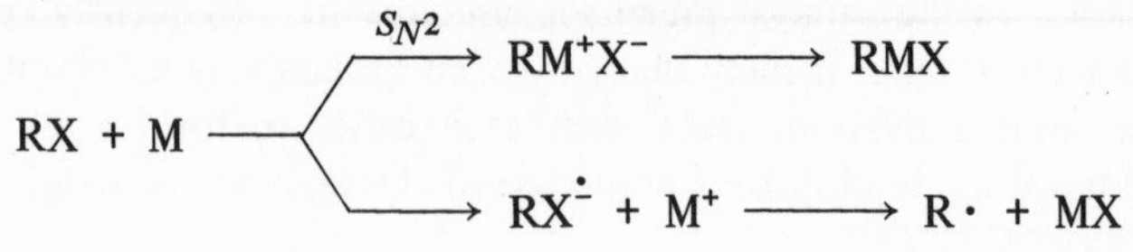

$$RX + M \begin{cases} \xrightarrow{S_N2} RM^+X^- \longrightarrow RMX \\ \\ \longrightarrow RX^{\cdot-} + M^+ \longrightarrow R\cdot + MX \end{cases}$$

Scheme 10.6

chain or a radical chain process depending on the fate of the radical R. All of these mechanisms have been observed for oxidative addition processes. Similarly, the reactions of metal (carbonyl) hydrides with olefins, alkyl halides, etc., appear to involve both one- and two-electron mechanisms depending on the substrate and conditions. It is worth emphasizing that up until quite recently most reactions of organotransition metal complexes were written as concerted, three-centre insertion processes, i.e. as "non-mechanisms". Recognition of the importance of stepwise, homolytic and heterolytic mechanisms, analogous to those observed in organic chemistry, represents an important breakthrough in understanding organometallic processes.

Recent mechanistic studies have also provided a better insight into the mechanism of hydrocarbon formation on metal surfaces. Metal carbenoid species appear to be key intermediates in these processes. Product formation involves surface polymerization of these intermediates and the product distribution is determined by the relative rates of propagation and termination. A better understanding of the factors influencing these rates would provide important clues for the development of more selective catalysts for lower olefin formation from syn gas.

In conclusion, we are of the opinion that syn gas chemistry has a bright future. We expect it to be widely applied in the synthesis of both bulk and fine chemicals, leading to fundamental, far-reaching changes in the structure of the chemical industry. It is very likely that this will involve, for a large part, the production and conversion of methanol. Methanol, produced on a very large scale at locations where primary raw materials are cheap, could be transported as a fuel-grade product to the point of use. Here it could be used as such, as a liquid fuel, or upgraded to chemical grade methanol and used as a petrochemical feedstock. In the final analysis, the key to success in syn gas/methanol-based chemistry probably lies in the development of fully integrated processes which maximize feedstock utilization.

References

1. D. R. Liu and M. S. Wrighton, *J. Am. Chem. Soc.*, **104**, 898 (1982).
2. J. W. Hershberger and J. K. Kochi, *J. Chem. Soc. Chem. Commun.*, 212 (1982).
3. A. Darchen, C. Mahé and H. Patin, *J. Chem. Soc. Chem. Commun.*, 243 (1982).

INDEX

Acetaldehyde
  aldol condensation, 164
  amidocarbonylation of, 170
  from ethanol dehydrogenation, 164
  from ethylene oxidation, 5, 141
  from formaldehyde homologation, 134
  hydrogenation of, 157
  from hydrogenolysis of acetic anhydride, 147
  from syn gas, 185
Acetic acid
  from butane oxidation, 7, 137, 141
  from ethylene, 5
  homologation of, 160
  from methanol carbonylation, 141–4
  from methyl formate isomerization, 137
  from syn gas, 185, 189
Acetic anhydride
  from ethylene, 5
  hydrogenolysis
  from methyl acetate carbonylation, 144
Acetone
  conversion to methylmethacrylate, 148
  from propylene, 6
Acetonitrile
  in amidocarbonylations, 171
  from syn gas and ammonia, 182
N-Acetylalanine
  from acetaldehyde amidocarbonylation, 170
N-Acetylamino acids ·
  via amidocarbonylation, 171
Acetyl chloride
  promotor in acetic anhydride hydrogenolysis, 148
Acetylene
  carbonylation of, 104
N-Acetylglycine
  from formaldehyde amidocarbonylation, 134
N-Acetylphenylalanine
  from benzyl chloride, 173

from styrene oxide amidocarbonylation, 170
Acrolein
  from propylene, 6
Acrylamide
  carbonylation of, 169
Acrylic acid
  from acetylene carbonylation, 106
  from oxidative carbonylation of ethylene, 111
  from propylene, 6
Acrylonitrile
  alkoxycarbonylation of, 110
  amidocarbonylation, 172
  methanol based route to, 182
  from propylene, 6
Activation of carbon monoxide, 31
Activation of hydrogen,
  mechanism of, 34
  by metal carbonyls, 35
N-Acylamino acids
  via amidocarbonylation, 170, 171
Acyloxymethylmanganese complex, 193
Acyloxymethylruthenium complex, 193
Adipic acid
  from butadiene hydroxycarbonylation, 110
  manufacture of, 8
Adiponitrile
  from butadiene, 6
Aldehydes
  from hydrogenolysis of acylmetal species, 141
  from organic halides and syn gas, 118
  reduction with $HCo(CO)_4$, 153
Aldol condensation, 134, 164
Alkoxycarbonylation, *see also* methoxy-carbonylation
  of acrylonitrile, 110
  of alkyl halides, 116
  of allylic halides, 116
  of aryl halides, 118